Advances in Soil Science

Advances in Soil Science

B.A. Stewart, Editor

Advances in Soil Science

Volume 7

Edited by B.A. Stewart

With Contributions by
D.W. Anderson, E. Bresler, W.W. Frye, L. Metzger, M.S. Smith, J.J. Varco,
J. Venkateswarlu, B. Yaron

With 39 Illustrations

Springer-Verlag
New York Berlin Heidelberg
London Paris Tokyo

B.A. Stewart
USDA Conservation & Production Research Laboratory
Bushland, Texas 79012, U.S.A.

ISSN: 0176-9340

Softcover reprint of the hardcover 1st edition 1987

Typeset by Asco Trade Typesetting Ltd., Hong Kong.
Printed and bound by R.R. Donnelley & Sons, Harrisonburg, VA.

9 8 7 6 5 4 3 2 1

ISBN-13: 978-1-4612-9157-2 e-ISBN-13: 978-1-4612-4790-6
DOI: 10.1007/ 978-1-4612-4790-6

Preface

The world population in 1950 was 2.5 billion and is more than 5 billion today. The agriculture community, however, has responded remarkably well to meeting the increased need for food and fiber. While the population growth during this period averaged almost 2% annually, the production of grain increased at an even faster rate. From 1950 to 1973, grain production increased 3.1% annually, but slowed to about 2% from 1973 to 1984. Therefore, as a whole, the world has more grain per capita available today than ever before. Several countries that were food importers just a few years ago are food exporters today. The world carry-over stocks today are the highest in years. While the major concern just a few years ago was whether the world could produce adequate food, the problem for many countries today is how to export their surpluses.

Although the world as a whole has surplus food supplies, there are millions of people without adequate food to exist, and there are additional millions who have a bare subsistence diet at best. The average daily calorie supply for the developed countries is more than 3,300 per person, while the average for the developing countries is only about 2,200. The major global food production problems have shifted from Asia to Africa, where malnutrition, poverty, and starvation are attracting world attention. Since 1967, per capita grain production in Africa has been declining, and even then, the grain production averaged only 180 k per person, which is considered the subsistence level. The average now is about 150 k, and in the drought years of 1983 and 1984, production dropped to less than 120 k per person. The challenge that Africa presents is enormous, because the resources available are limited, particularly in the Sahel region.

The tremendous growth rate in food production has been accompanied by an even greater growth rate of energy use by agriculture. While grain production increased from 624 million metric tons in 1950 to 1,661 million metric tons in 1985, energy use in terms of oil equivalent increased from 276 million barrels in 1950 to 1,903 million barrels in 1985. The use of energy for fertilizer and irrigation fuel increased nearly 10-fold during this period. Since the share of total oil used by agriculture is relatively small, there is no immediate concern of an insufficient supply. However, relative energy costs are expected to rise in future years, and this will make food costs higher if agriculture continues its dependence on oil. Although world agriculture production continues to increase as energy use increases, the amount of food produced for each additional unit of energy used declines steadily. Therefore, there is a growing realization that agriculture production systems must be developed that are less energy dependent.

An equal concern is developing about the effect that agriculture production systems are having on the soil resource base. Deforestation is a major problem in many developing countries. The primary cause of deforestation in these countries is the critical need of firewood to supply a rapidly growing population. The soil resource base is also being seriously damaged in many regions by wind and water erosion, salinization, compaction, increased acidity, and depletion of nutrients.

These concerns are resulting in a shift in research objectives. Future efforts will be focused more on the development of cropping systems that are more resource efficient. The relationships between input costs and yield will be carefully scrutinized. The need to optimize crop production while conserving the resource base has never been greater.

This series, *Advances in Soil Science*, was established to provide a forum for leading scientists to analyze and summarize the available scientific information on a subject, assessing its importance and identifying additional needs. But most importantly, the contributors are asked to develop and identify principles that have practical applications to both developing and developed agricultures. It is not the purpose of the series to report new research results, because there are many excellent scientific journals for that need. Communications in scientific journals, however, are generally restricted to short and technical presentations. Therefore, *Advances in Soil Science* fills a gap between the scientific journals and the comprehensive reference books in which scientists can delve in depth on a particular subject. The quick acceptance of the series by both authors and readers confirms our perception that a need did exist for a medium to publish reviews relating to soil science. I wish to acknowledge the authors for their excellent contributions and cooperation. I should also like to thank the members of the Editorial Board for their assistance in selecting such competent and pleasant authors with whom I had the pleasure of working. Lastly, and most importantly, I want to thank the readers for their acceptance and use of *Advances in Soil Science*.

B.A. Stewart

Contents

Contributors

D.W. Anderson, Saskatchewan Institute of Pedology, University of Saskatchewan, Saskatoon S7N 0W0, Canada

E. Bresler, Institute of Soils and Water, The Volcani Center, Bet-Dagan 50-250, Israel

W.W. Frye, Department of Agronomy, University of Kentucky, Lexington, Kentucky 40546, U.S.A.

L. Metzger, Institute of Soils and Water, The Volcani Center, Bet-Dagan 50-250, Israel

M.S. Smith, Department of Agronomy, University of Kentucky, Lexington, Kentucky 40546, U.S.A.

J.J. Varco, Department of Agronomy, University of Kentucky, Lexington, Kentucky 40546, U.S.A.

J. Venkateswarlu, Department of Agriculture and Cooperation, Ministry of Agriculture, New Delhi 110001, India

B. Yaron, Institute of Soils and Water, The Volcani Center, Bet-Dagan 50-250, Israel

Modeling of Flow, Transport, and Crop Yield in Spatially Variable Fields

Eshel Bresler*

I. Introduction

The traditional approach of modeling processes occurring in crop and soil systems is to apply macroscopic laws derived from laboratory scale, or small-field scale, to large-scale fields. Applications of such models to actual field conditions were supported by the assumption that the field can be regarded

* Institute of Soils and Water, The Volcani Center, Bet Dagan 50250, Israel.

as a homogeneous or equivalent-to-homogeneous medium characterized by equivalent or effective soil properties that, in turn, are measured by sampling over a few locations in the particular field, and determined subsequently by an averaging procedure. Such an approach assumes that agreement between the outcome of the model for the fictitious homogeneous field and the process takes place in the actual, spatially variable field. It raises, however, the question of whether or not the variables (solute concentration, water content fluxes, crop yield, etc.), as depicted by the models, equal the average of the same variables prevailing in the field, and, if so, how to average the soil properties over the field to get such an equivalence. Also, is it sufficient to determine the average of these quantities (concentration, water flow properties, crop yield), or it is necessary to determine the confidence interval of the model results?

A rational approach to field variability is that of stochastic modeling. Soil properties and, therefore, the variables' outcome from model, are subjected to uncertainty and are regarded as random variables that can be defined in terms of their statistical moments. The actual field is viewed as a realization of all the various possible fields that have the same properties as the given field. Assuming stationarity and adopting the ergodic hypothesis, the statistical ensemble average and space average can be interchanged. Hence, the expectation of any model variable is equal to its arithmetic mean over the field, and its variance is a measure of the deviation of the variable from its mean at various points in the field. The stochastic models are therefore employed to predict the dependence of the statistical moments of the variables like crop yield, concentration, water content, and so forth, on those of soil properties for a given set of man-controlled variables (initial and boundary conditions) such as irrigation water quantity, salt concentration of the irrigation water, and so forth. These statistical moments exhaust the information that the modeler would need in any conceivable application in agriculture, such as moments of crop yield or farm income as related to statistical moments of water content and salt concentration, which in turn reflect their distribution in the field.

Models aimed at solving the problem of flow and transport and their relationships to crop yield in a homogeneous soil are quite difficult to apply because of the strong nonlinearities of the differential equations. Also, an accurate description of the soil properties on which the model results depend is difficult to achieve and requires time-consuming measurements. However, since the farmer's interest lies in a few statistical averages over the field, rather than a detailed description of the variables of interest in space and time, the estimation of space averages of water content, soil salinity, and even crop yield, may be relatively simple in a few cases, and, paradoxically, even simpler than solution of the problem in a homogeneous field. The reason may be that in the process of averaging over the field, mutual cancellation of errors may occur. In addition, the variance due to spatial variability may, in many cases, be much larger than the variance related to the approximations incurred by the model, rendering the use of more accurate models irrelevant. Hence, in spatially variable fields, stochastic modeling may represent realistically the actual field phenomena, and may provide the main statistical moments—mean and variance—by simplified models.

These general ideas are expressed in the models that are described in this chapter for a few processes of water flow and solute transport encountered in many applications in agriculture: (1) Nonsteady infiltration caused by application of a given constant water flux at the surface (unless ponding occurs), and with a given salt concentration, into soil of initially constant water content and salt concentration; (2) subsequent (postinfiltration) redistribution and salt transport under these conditions; (3) piston-type salt transport under steady gravitational flow conditions using closed-form analytical solution, and dispersive salt transport under identical steady gravitational conditions; (4) modeling of crop yield for a spatially variable field, using the crop yield models discussed in Bresler et al., (1982) and the Mitcherlich expression; and (5) simple economic irrigation model for a spatially variable soil. This last model is based on the work done by this author and his colleagues, Eli Feinerman and Gedeon Dagan (see Feinerman et al., 1985). The rest of the models are those of Gedeon Dagan and this author, the details of which are given in Dagan and Bresler (1979, 1983) and Bresler and Dagan (1979, 1981, 1983a, 1983b).

II. Infiltration and Redistribution

The traditional approach (e.g., Hanks et al., 1969) is to model water flow by partial differential equations (e.g., Equation [1]) subject to appropriate boundaries and initial conditions (e.g., Equations [2]–[5]), while soil hydraulic properties are represented by numerical tables or empirical relationships between hydraulic parameters (e.g., Equations [6] and [7]). In the particular problem of infiltration and redistribution, water is applied for a certain time on the soil surface and the solution describes the flow of water in the soil profile as a function of time (t) and depth (z). The validation of the equations and their solutions has generally been carried out under controlled laboratory conditions by using soil columns. The solution of the problem for a homogeneous soil column is quite difficult because of the strong nonlinearity of the differential equation (Equation [1]), and the time required for accurate measurements of the soil hydraulic properties (Equations [6] and [7]). These difficulties are generally not amplified in a field of spatially variable properties, because our interest is in a few statistical averages over the horizontal plane as a function of z and t over the field, rather than a detailed description of water content, for instance, as a function of the space variables and time. Paradoxically, the estimating of the average over the horizontal plane may be even simpler than solving the flow problem in a homogeneous column. The reason is that flow parameters in a model of a spatially variable field are subjected to two kinds of errors: those related to the uncertainty of the values of soil properties at different locations, and those related to the use of inaccurate, approximate flow models, the latter being dependent on the first. In the process of averaging over the field, mutual cancellation of errors may occur (Bresler and Dagan, 1983a, 1983b). Furthermore, the variance due to spatial variability may be much larger than that related to model approximation, rendering the use of accurate models irrelevant (the same type of problem is encountered in various branches of physics, like statistical mechanics, in which simplified microscopic models lead to quite accurate macroscopic relationships). Hence, in spatially variable fields, stochastic modeling may represent more realistically the actual flow phenomena on the one hand, any may provide the main statistical moments (mean, variances) by simplified flow models on the other hand. These general ideas are applied here to the two processes: infiltration into soil of initially constant water content, and subsequent redistribution.

A. Mathematical Statement of Flow Problem

Soil properties are assumed to be uniform along the vertical (but to vary in the horizontal) plane, and flow is vertical. The soil lies beneath the horizontal plane x, y with z, a vertical coordinate, directed downward. Let q be the vertical flux, θ the soil moisture, ψ the suction head, and K the hydraulic conductivity. The one-dimensional Richard's equation for θ is:

$$\frac{\partial \theta}{\partial t} + \frac{\partial}{\partial z}\left(K \frac{d\psi}{d\theta}\frac{\partial \theta}{\partial z}\right) + \frac{\partial K}{\partial z} = 0 \qquad [1]$$

The boundary and initial conditions are as follows:

$$\theta = \theta_n \qquad t = 0 \qquad z > 0 \qquad [2]$$

$$q = q_0 \qquad 0 < t < t_i \qquad z = 0 \text{ for } \theta(0, t) < \theta_s) \qquad [3]$$

$$q = 0 \qquad t > t_i \qquad z = 0 \qquad [4]$$

where θ_n ($\geqslant \theta_r$) is the constant, initial moisture content, q_0, is the given constant flux on the surface, and t_i is the infiltration time. If $\theta(0, t)$ reaches the saturated value θ_s for $t_p < t < t_i$, Equation [3] is replaced by

$$\theta = \theta_s \qquad t_p < t < t_i \qquad z = 0 \qquad [5]$$

where t_p is defined as ponding time, that is, the time at which ponding occurs and persists until t_i. Equation [5] neglects the effect of water accumulation on the surface and the eventual positive head or runoff caused by it.

The $K(\theta)$ and $\psi(\theta)$ relationships needed in order to solve Equation [1] are often determined empirically. Many analytical representations that fit measurements with various degrees of success have been suggested in the past. The type of relationships of Brooks and Corey (1964) is the one adopted here, as

$$\frac{K(\psi)}{K_s} = \left(\frac{\psi_w}{\psi}\right)^{\eta} \qquad [6]$$

$$S = \frac{\theta - \theta_r}{\theta_s - \theta_r} = \left(\frac{\psi_w}{\psi}\right)^{\beta} \qquad [7]$$

Here, K_s is the hydraulic conductivity at saturation, ψ_w is the air entry value, S is the reduced water content (saturation), θ_s and θ_r are moisture contents for saturation and for $K \to 0$, respectively, and η and β are constant, empirical coefficients ($\eta \sim 2.0$–3.5, $\beta \sim 0.25$–0.5, $\eta/\beta \sim 7.2$). Thus, the soil hydraulic properties are characterized by the six constants $\theta_s, \theta_r, \psi_w, K_s, \eta$, and β. Other types of relationships could be adopted as well, but the influence of their precise nature upon statistical moments is quite limited. Note that hysteresis in the relationships of Equations [6] and [7] is neglected here.

The solution of Equation [1] associated with the "constituitive" relationships of Equations [6] and [7] and boundary and initial conditions in Equation [2]–[5], has been the subject of numerous studies aiming at obtaining approximate analytical expressions or numerical solutions for $\theta(z, t; q_0, \theta_n, t_i, \theta_s, \theta_r, \psi_w, \eta, \beta, K_s)$. The main purpose here, however, is to solve Equation [1] for a field of hydraulic properties that vary in space.

B. Outline of Methodology of Stochastic Flow Problem

The six parameters of Equations [6] and [7] are the coordinates functions of x and y in the plane. Since they are determined by measurements at only a few locations, which are error prone, and since they generally vary in an irregular fashion in space, we regard them as random variables of x and y. A considerable simplification is achieved by assuming that only K_s is spatially variable, while the five other parameters $(\theta_r, \theta_s, \psi_w, \eta, \beta)$ are constant. The justification for this assumption is that K_s may change considerably over the field, while the other parameters vary within much narrower limits. Previous statistical analysis (Russo and Bresler, 1982a) indicated that the impact of the variability of these parameters is indeed limited. The type of analysis, however, can be easily generalized for more parameters, including water application rate q_0, which also has been taken here as a deterministic constant (the case of random q_0 will be considered later in relation to salt transport). The assumption of vertical, one-dimensional flow implies neglect of horizontal velocity components in comparison with the vertical one. The horizontal components arise due to soil heterogeneity, but their magnitude depends essentially on the correlation scale in the horizontal plane. If this scale is much larger than the vertical length scale associated with the head gradient, the horizontal gradients are negligible (see Russo and Bresler, 1982b), as assumed here.

Under these conditions, the space variability is expressed mathematically with the aid of the joint probability density function (PDF) $f_k(K_{s1}, K_{s2}, \ldots, K_{sn})$

of K_s at a set of n points of coordinates r_i (x_i, y_i), $(i = 1, 2, \ldots, n)$. Since there is, in applications, only one field, the space variability in this single realization can be used to derive f_k only under the stationary hypothesis. It will be assumed, in line with previous studies and field measurements, that K_s is weakly stationary, that is, the expectation of K_s does not depend on r_i, and it is constant, so that the two-point correlation depends only on the distance $r_{12} = |r_1 - r_2|$ between the points. In this case, the statistical structure of K_s is determined, up to second-order moments, by the probability density function $f_k(K_s)$, irrespective of r_i, and by the autocovariance.

Since θ (or S), ψ, and K depend on K_s, via Equations [6] and [7], they are random variables as well. Furthermore, Equation [1] is a stochastic partial differential equation, subjected to the deterministic boundary and initial conditions of Equations [2]–[5] (as mentioned before, incorporation of randomness in the latter equations is possible).

The problem of infiltration and redistribution in a spatially variable field can now be stated as follows: determine the statistical structure of the random variables, θ (or S) and the related variables, ψ and K, which satisfy Equation [1], the relationships of Equations [6] and [7], and conditions in Equations [2]–[5] for given stationary K_s. If the scope is limited to a few statistical moments of the dependent variables only, namely, to their unconditional expectation $E(S)$ and variance (σ_S^2), then

$$E[S(z,t)] = \int_0^\infty S(z,t;K_s)f_k(K_s)\,dK_s \qquad [8]$$

and

$$\sigma_S^2(z,t) = \int_0^\infty \{S(z,t;K_s) - E[S(z,t)\}^2 f_k(K_s)\,dK_s \qquad [9]$$

and similarly for ψ and K. Again, the assumptions underlying Equations [8] and [9] are that all soil and flow parameters, other than K_s, are deterministic and constant, so that variables like S, K, and ψ change with x,y only through K_s. It is seen, therefore, that once the relationship between S and K_s is determined by solution of Equations [1]–[7], the moments of interest are obtained by one quadrature over the PDF of K_s. In this respect, three points are of interest:

1. The correlation structure of K_s, expressed by its autocovariance, does not enter into the computation of Equations [8] and [9]. This quantity may be of interest in some applications, and its computation, following this methodology, is straightforward.
2. The numerical evaluation of the integrals in Equations [8] and [9] can be carried out by replacing them by sums over equal classes of the cumulative PDF

$$P_k(K_s) = \int_0^{K_s} f_k(K_s)\,dK_s \qquad [10]$$

This is more economical, in terms of computing time, than a straightforward Monte Carlo simulation (e.g., Haan, 1977), which is not discussed in this article, of K_s and subsequent evaluation of Equations [8] and [9].

3. Under the stationarity hypotheses adopted here, $E[S(z,t)]$, for instance, represents the average saturation in the field in the plane at depth z and at time t, that is,

$$E[S(z,t)] \cong \iint S(x,y;z,t)\,dx\,dy \qquad [11]$$

and similarly for σ_S^2. The equivalence of Equations [8] and [11] is underlain by the requirement that the integral scale of K_s is much smaller than the length scale characterizing the extent of the field in the x,y plane. This condition is supposed to be met here. (For a discussion of the estimation errors incurred by the finiteness of the field, see, for example, Russo and Bresler [1982b].)

In line with many studies of the PDF of K_s, it is assumed that it is related to scaling parameter δ, which is, in turn, lognormal, that is,

$$Y = \ln \delta \qquad [12]$$

and its PDF f_Y is given by

$$f_Y(Y) = \frac{1}{(2\pi)^{1/2}\sigma_Y} \exp\left[-\frac{(Y - m_Y)^2}{2\sigma_Y^2}\right] \qquad [13]$$

The relationship between K_s and δ is obtained as follows: From similarity considerations about the microscopic flow equation, one gets (Miller and Miller, 1956)

$$K_s = \lambda d^2 \quad \text{or} \quad K_s^{1/2} = \lambda^{1/2} d \quad \text{and} \quad E[K_s^{1/2}] = \lambda^{1/2} E[d]$$

where d is the characteristic length of the pores and λ is the proportionality constant. If we defined δ by $\delta = d/E[d]$ then

$$\frac{K_s^{1/2}}{E[K_s^{1/2}]} = \frac{\lambda^{1/2} d}{\lambda^{1/2} E[d]} = \frac{K_s^{1/2}}{K_s^*} = \delta$$

or

$$K_s/K_s^* = \delta^2 \qquad [14]$$

where

$$K_s^* = [E(K_s^{1/2})]^2 = \left[\int_0^\infty K_s^{1/2} f(K_s)\,dK_s\right]^2 \qquad [15]$$

Hence, the PDF of K_s is entirely defined by the three parameters $E(Y) = m_Y$, σ_Y, and K_s^*. Once they are known, equations like [8] and [9] can be easily evaluated by using the transformations

$$K_s = K_s^* \exp(2Y) \quad [16]$$

$$f_k(K_s)\,dK_s = f_Y(Y)\,dY \quad [17]$$

It is seen, therefore, that once the deterministic parameters θ_s, θ_r, ψ_w, β, η, m_Y, σ_Y, K_s^* (characterizing the field) and q_0, t_i, θ_n (characterizing the flow) are given, the calculation of the average (Equation [8]) and variance (Equation [9]) boils down to solving first Equations [1]–[7] to determine $S(z, t; K_s)$ for an arbitrary K_s and, subsequently, integrating over Y with the aid of Equations [16] and [17].

The solution of Equation [1] subject to Equation [2] through [5] for given K_s is quite difficult and can be carried out by numerical methods, some of which are described in Bresler et al., (1982). The objective of this section is to demonstrate that one can obtain quite accurate estimates of expectation and variances for a spatially variable field, even if a simple approximate solution to Equations [1] through [5] is used.

C. Approximate Water Flow Model

The simplest approach applicable to the constant surface flux conditions (Equation [3]) is the one based on the motion of water content front (Green and Ampt approach) following the study of Mein and Larson (1973). The basic approach is to replace the differential equation (Equation [1]) by its integrated expression between the soil surface and the front. The complete development of the approximate front model from basic principles for both infiltration and redistribution processes is described in detail by Dagan and Bresler (1983) and is not reported here. Only the assumptions of the approach, part of which has been described already, and the working equations are presented here.

1. Summary of Basic Assumptions

The following assumptions are adopted (not necessarily in order of their importance):

1. The flow direction is vertical.
2. The soil hydraulic properties do not change along any vertical profile, but vary considerably in the horizontal plane. This variation of the properties can be described in terms of a probability distribution function.
3. The hydraulic conductivity-water content relationship $K(\theta)$ has a simple analytical structure (Equations [6] and [7]).
4. The randomness of $K(\theta)$ stems from the stochastic nature of the saturated conductivity K_s, which depends quadratically on a scaling parameter δ (Equation [14]).
5. The scaling parameter (δ) has a lognormal distribution, that is, Y in Equation 12 ($Y = \ln \delta$) is normal Equation [13] with mean m_Y and variance σ_Y^2.

6. Unsteady water flow is generated by deterministic recharge applied on the surface at a rate (q_0) during the infiltration time (t_i) (see Equation [3]).
7. Water content is initially deterministic and homogeneous throughout the entire field.
8. A piston flow-type profile is assumed at any point in the field during both infiltration (Equations [3] and [5]) and redistribution (Equation [4]), processes.

Under these conditions, the flux (q) and the water content (θ) are step functions along vertical lines, with constant values depending on soil properties, for profiles shallower than the wetting front ($z < L$), and with values equal to the initial ones, for $z > L$. Hence, at a given time (t) and depth (z), the magnitude of $q(z, t)$, $\theta(z, t)$, the other related flow variables, and the wetting depth (L) may vary considerably in the horizontal plane due to the spatial variability of the soil hydraulic properties.

2. Summary of Model Solution

The main assumption is the replacement of the actual water content profile by a fictitious uniform profile extending from the soil surface to an equivalent wetting front at depth $L(t)$. Darcy's law is satisfied in the average over the profile and the water mass conservation is obeyed exactly. The steps for the solution of one infiltration-redistribution cycle in the approximate deterministic problem are as follows:

Given the input parameters K_s, η, β, ψ_w, θ_r, θ_s, θ_n, q_0, and t_i, calculate the dimensionless variables

$$\tilde{t} = \frac{t q_0(\eta - 1)}{\psi_w(\theta_s - \theta_r)}; \qquad \mu = 1 - \frac{1 - \beta}{\eta}$$
$$\tilde{K} = \frac{\bar{K}}{q_0}; \qquad \tilde{L} = \frac{L(\eta - 1)}{\psi_w} \qquad [18]$$

and the ponding time (t_p)

$$\tilde{t}_p = \frac{\tilde{K}_s[1 - (\tilde{K}_n/\tilde{K}_s)^{(1-1/\eta)} - S_n + S_n(\tilde{K}_n/\tilde{K}_s)^{(1-1/\eta)}]}{(1 - \tilde{K}_n)(1 - \tilde{K}_s)} \qquad [19]$$

where $\bar{K} = K(\bar{\theta}) = K(\bar{S})$ is average K over the wet profile $0 < z \leqslant L$ and $K_n = K(\theta_n)$, $S_n = S(\theta_n)$ are the initial, constant values of K and S for $z > L$. If the ponding time (t_p) is larger than the infiltration time (t_i), then $K(t)$ is calculated throughout the entire infiltration stage from

$$\frac{\tilde{K}}{\tilde{K}_s} + [(\tilde{K}/\tilde{K}_s)^{\mu} - \tilde{K}_n^{1-1/\eta} K^{\beta/\eta} \tilde{K}_s^{-\mu} - S_n(\tilde{K}/\tilde{K}_s)^{(1-1/\eta)}$$
$$+ S_n(\tilde{K}_n/\tilde{K}_s)^{(1-1/\eta)}][(1 - \tilde{K}_n)\tilde{t}]^{-1} = \frac{1}{\tilde{K}_s} \qquad [20]$$

Subsequently, the saturation $\bar{S}(t)$ (Equation [7]) and the depth $L(t)$ are derived from

$$\bar{S} = \left(\frac{\tilde{K}}{\tilde{K}_s}\right)^{\beta/\eta} \qquad [21]$$

and

$$L = V/(\theta - \theta_n) = (q_0 - K_n)t/[(\theta_s - \theta_r)(\bar{S} - S_n)] \qquad [22]$$

If, on the other hand, t_p as calculated from Equation [19] is smaller than t_i, then Equation [20] is used as long as $0 < t < t_p$. When the soil is saturated during infiltration (i.e., for $t_p < t < t_i$), the value of $V(t)$, which is the volume of water per unit area of soil accumulated in the profile from $t = 0$ to any t that is,

$$V(t) = \int_0^L [\theta(z,t) - \theta_n]\,dz = [\bar{\theta}(t) - \theta_n]L(t), \qquad [23]$$

is calculated from

$$\tilde{V} - \tilde{V}_p - \alpha \ln \frac{\alpha + \tilde{V}}{\alpha + \tilde{V}_p} = (\tilde{t} - \tilde{t}_p)(\tilde{K}_s - \tilde{K}_n)$$

$$\tilde{V} = \frac{V(\eta - 1)}{\psi_s(\theta_s - \theta_r)}; \qquad \alpha = \frac{(1 - S_n)(1 - S_n^{(\eta-1)/\beta)})}{1 - (\tilde{K}_n/\tilde{K}_s)} \qquad [24]$$

Note that from the same conditions ($t_p < t < t_i$) $\bar{S} = 1$, L is obtained from Equation [22] while surface flux $q(0, t)$ is calculated from

$$q(0,t) = \frac{d\tilde{V}}{d\tilde{t}} = K_s + \frac{K_s \psi_w(\theta_s - \theta_n)}{(\eta - 1)V}; \; t > \frac{\psi_w(\theta_s - \theta_n)}{(\eta - 1)(q_0 - K_s)} \qquad [25]$$

During redistribution (for $t > t_i$), values of $K(t)$ are obtained by numerical integration of

$$\frac{d\bar{K}}{dt} = -\frac{\eta}{\beta} \frac{\bar{K}^{1-\beta/\eta} K_s^{\beta/\eta}[(\bar{K}/K_s)^{\beta/\eta} - S_n]}{V_0 - K_n t} \left[\bar{K} - K_n + \frac{\psi_w}{\eta - 1} \times \frac{(\bar{K}^{(1-1/\eta)} K_s^{1/\eta} - K_n S_n^{-1/\eta})(K^{\beta/\eta} K_s^{-\beta/\eta} - S_n)}{V_0 - K_n t} \right] \qquad [26]$$

$\bar{S}$ is subsequently derived by Equation [21], while L results from Equation [23] and from

$$V(t) = V_0 - K_n t \qquad t > 0 \qquad [27]$$

where V_0 is water volume in the profile at the termination of the infiltration process ($t = t_i$), but for the sake of convenience, $t = 0$ is taken at the beginning of redistribution.

D. Computations of Statistical Moments of Flow Variables

To calculate the mean and variances of the various flow variables, the integrals of Equations [8] and [9] have to be evaluated. Although for some specific flow conditions, such as steady gravitational flow for solute transport (see later), closed form analytical evaluations are possible, numerical computations have to be performed for the more general flow conditions considered here. Toward this goal, the well-known Monte Carlo simulation method can be used to obtain the appropriate PDF of K_s with known values of σ_Y and m_Y. An alternative method, which has been adopted for this model, is to divide the cumulative PDF of Y,

$$P(Y) = \int_{-\infty}^{Y} f(Y)\,dY \qquad [28]$$

into N equal classes ($i = 1, 2, \ldots, N$), to calculate K_s for each class from

$$K_s^i = K_s^* \exp(2Y_i) \qquad [29]$$

and to evaluate numerically the integrals in Equations [8] and [9] with computed values of $S(z, t; K_s^i)$. This method is more economical in terms of computing time than the conventional Monte Carlo simulation of K_s and subsequent evaluation of the integrals.

The computation of K_s^i for $i = 1, 2, \ldots, N$ is carried out as follows.

1. The area $P(Z)$ under the standard normal density curve $f(Z)$ for Z: $N[0, 1]$ has been divided into N equal classes, and the value of Z_i in each class i has been recorded.
2. By using the inferred values of σ_Y and m_Y, the values of Y_i for a class of soil are obtained from

$$Y_i = Z_i \sigma_Y + m_Y \qquad [30]$$

These Y_i values are then substituted into Equation [29] and K_s^i for $i = 1, 2, \ldots, N$ are obtained.

After establishing N values of Y or their equivalent K_s, Equation [1] and [2]–[5] can be solved numerically (e.g., Hanks et al., 1969) or approximately to obtain any flow variable $\mu = S, \psi$, or K. The computed $\mu(z, t; Y)$ values are then substituted into equations similar to Equations [8] and [9] to calculate the statistical moments of μ by numerical evaluation of the appropriate integrals. The procedure for computing the statistical moments of μ with the eight parameters characterizing the field ($m_Y, \sigma_Y, K_s^*, \eta, \beta, \theta_s, \theta_r, \psi_w$) and the three parameters characterizing the flow (q_0, t_i, θ_n) for the approximate and numerical methods are given next.

1. Computations by Approximate Model

The actual computations involve the following steps:

1. Substitute in Equation [30] the given values of K_s^*, σ_Y, m_Y, and Z_i for $i = 1, 2, \ldots, N$ to obtain Y_i, and then calculate K_s^i from Equation [29].

2. Set $i = 1$ and define $\tilde{t}$, μ, and $\tilde{K}^i$ in Equation [18] from the input values of η, β, θ_s, θ_r, ψ_w, q_0, t_i (the starting time of redistribution), z, and t and from the calculated K_s^i.
3. Compute $\tilde{t}_p^i$ from Equation [19] and compare $\tilde{t}_p$ and $\tilde{t}_i$ with t. If $\tilde{t}_i > \tilde{t} > \tilde{t}_p$, set $S_i = 1$, $\tilde{K}_i = \tilde{K}_s^i$, $\tilde{V}_p = \tilde{t}_p$ (because $V_p = t_p q_0$ and because of the definition in Equation [18]) and calculate $\tilde{V}_i$ from Equation [24] by using Newton's iterative method (see Abramowitz and Segun, 1964), $V_i(t)$ from the definition of $\tilde{V}$ (Equation [24]) and $L_i(t)$ from

$$L_i(t) = V_i(t)/[(\theta_s - \theta_r)(1.0 - S_n)] \qquad [31]$$

If ($\tilde{t} < \tilde{t}_p$), set $\tilde{V}_i = \tilde{t}$ (because $V = tq_0$ and calculate $K_i(t)$ from Equation [20] again by Newton's iterative method. Then compute $S_i(t)$ from Equation [21], $\tilde{V}_i(t)$ from V_i, and $L_i(t)$ from Equation [31]. If ($\tilde{t} > \tilde{t}_i$), solve for $K_i(t)$ the differential equation (Equation [26]) with the aid of the Runge-Kutta method (see Abramowitz and Segun, 1964), and again compute $S_i(t)$ from Equation [21] and $L_i(t)$ from Equation [31].
4. Compare $L_i(t)$ with the input value of z. If $L_i(t) > z$, compute $G_i(z, t) = dH/dz$ and $q_i(t)$ from

$$G_i(z, t) = 1 + \frac{\psi_w}{(\eta - 1)L_i(t)} \cdot [S_i(t)^{-1/\beta} - q_n(q_0/K_i(t))S_i(t)^{-1/\beta}] \qquad [32]$$

$$q_i(z, t) = K_i(t)G_i(z, t) \qquad [33]$$

where $K_i(z, t) = \tilde{K}_i(t)q_0$. If, on the other hand, ($L_t(t) < z$), then $\bar{S}_i(z, t) = S_n$, $K_i(z, t) = S_n^{\eta/\beta}K_s^i$, $G_i(z, t) = 1.0$, $q_i(z, t) = K_i(z, t)$.
5. Set $i = 2, 3, \ldots, N$, and repeat Steps 2–4.
6. Once $u_i(z, t)$ has been determined for the whole set of N values of the variable u, (where u stands for any variable like S, c etc) various moments of u, such as averages, variances, skewness, are computed from

$$\bar{u}(z, t) = \frac{1}{N} \sum_{i=1}^{N} u_i(z, t) \qquad [34]$$

$$\sigma_u^2(z, t) = \frac{1}{N} \sum_{i=1}^{N} [u_i(z, t) - \bar{u}(z, t)]^2 \qquad [35]$$

$$\gamma_u(z, t) = \frac{1}{N} \Sigma [u_i(z, t) - \bar{u}(z, t)]^3 \qquad [36]$$

7. Change z and/or t and repeat Steps 2–6 above.

2. Computation of Statistical Moments by Numerical Method

Finite difference equations to approximate the partial differential equation (Equation [1]), the initial conditions (Equation [2]), and the boundary conditions (Equation [3]–[5]) are first formulated and then solved numerically. The actual boundary conditions assigned at the surface at $t > 0$ are, therefore

$$q(0,t) = -K(0,t)\left[\frac{\partial\psi(0,t)}{\partial z} - 1\right] \leqslant q_0 \qquad [37]$$

$$\psi_w \leqslant \psi \leqslant \psi_w S_n^{-1/\beta} \qquad [38]$$

during infiltration, while $q_0 = 0$ during redistribution. Actually, a solution is sought by maximizing the value of $q(0,t)$ subject to the requirements in Equations [37] and [38] during infiltration. Here, during the first iteration in each time step, the surface is treated as a prescribed flux boundary and is assigned an arbitrary fraction of the potential flux q_0. If the computed value of $\psi(0,t)$ satisfies the requirement in Equation [38], then the value of $q(0,t)$ is increased. If the value of ψ lies outside the limits specified by Equation [38], then during the next iteration, the surface is treated as a prescribed pressure head with $\psi = \psi_w$. This situation is maintained as long as Equation [37] is satisfied. If at a stage of the computation the calculated flux exceeds the potential flux so that Equation [37] is not satisfied, potential flux is assigned to the surface and is again treated as a prescribed flux boundary. The iterative process continues until convergence is attained. Numerical methods of similar type were validated under various conditions of infiltration and redistribution under laboratory columns and fields (e.g., Hanks et al., 1969; Bresler and Laufer, 1974; Feddes et al., 1974; and Bresler et al., 1979).

The soil moisture retention curve that is adapted for the numerical computations is that of Equation [7], with the same values of β, ψ_w, θ_s, and θ_r as for the approximate simulated model. Hysteresis in the $\theta(\psi)$ relationship was not considered in the computations. The hydraulic conductivity function (Equation [6]) is also calculated as in the approximate model with the same deterministic η, K_s^*, m_Y, and σ_Y and the same values of K_s^i, $i = 1, 2, \ldots, N$. The computed results of the flow variables $u_i(z, t, Y_i)$, $i = 1, 2, \ldots, N$, are again substituted into Equations [34], [35], and [36] to calculate the three central statistical moments of $u = s, \psi$, and K.

E. Model's Results: Example for a Highly Variable Soil

To demonstrate best the stochastic model for infiltration and redistribution, a soil with high variation of K_s should be chosen. For this purpose, the values of the distribution parameters m_Y and σ_Y for the Panoche soil of Warrick et al. (1977) have been adopted. For the normal distribution of $Y = \ln \delta$, the values of the statistical distribution parameters are, for Panoche soil,

$$m_Y = -0.616 \qquad \sigma_y = 1.16 \qquad [39]$$

This Panoche soil is of low average conductivity and with quite large variability. The functions $K(\psi)$ (Equation [6]), and $S(\psi)$ or $\theta(\psi)$ (Equation [7]) have been adopted with values of $\eta = 2.59$, $\beta = 0.36$ (Bresler et al., 1978), $\theta_r = 0.05$ or 0.01, and $\psi_w = 15$ cm. Also, $\theta_s = 0.43$ and $K_s^* = 0.22$ cm/hr (Warrick et al., 1977) have been used. With these values of the eight deterministic parameters m_Y, σ_Y, K_s^*, θ_s, θ_r, β, η, and ψ_w (characterizing the field), one is able to compute

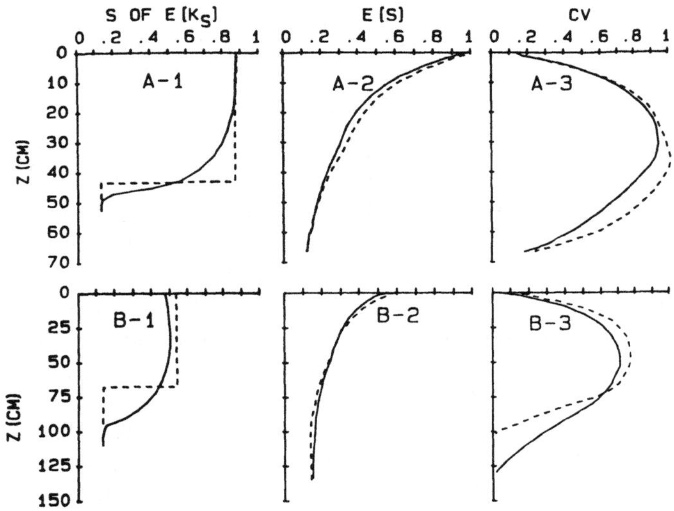

Figure 1. Computed profiles of degree of saturation (S) as a function of depth (z) for $\theta_n = 0.1$, $q_0 = 0.5$ cm/hr, after: A, $t_i = 2$ days of infiltration; B, $t - t_i = 5$ days of redistribution. 1, Computations using deterministic value of $E(K_s)$; 2, expected value of S ($E[S]$); 3, coefficient of variation of S (CV). Solid lines were computed with the numerical model; dashed lines with the "approximate" model.

the desired moments of the flow variables ($\theta(z,t)$ or $S(z,t)$, $\psi(z,t)$, etc.) for various combinations of the initial and boundary parameters (q_0, t_i, and θ_n) characterizing the flow.

The numerical computations (Hanks et al., 1969) were carried out with 90 depth increments. The size of each increment was constant with depth, but changed according to the value of K_s. For example, it ranged from 0.05 to 2 cm for $q_0 = 0.5$ and from 0.1 to 4 cm for $q_0 = 6.5$. The minimum value of Δt was chosen to be 0.02 hr.

Computations of mean and coefficient of variation (CV) of the saturation S (Equation [7]), for $\theta_n = 0.1$, $q_0 = 0.5$ cm/hr and $t_i = 2$ days at $t = 2$ days and at $t = 7$ days (after 5 days of redistribution), are demonstrated in Figure 1. Also given in Figure 1 (A-1, B-1) are deterministic water content profiles in terms of S calculated with deterministic K_s corresponding to its arithmetic average 0.93 cm/hr. This demonstrates the capability of the approximate model to simulate deterministic field conditions during infiltration and redistribution processes. It can be seen that the agreement between numerical and approximate models is not so close. This suggests that the approximate piston-type solution for water content (and S) is not accurate if a particular deterministic value is taken for K_s (and the other parameters). The picture is improved markedly when the statistical moments of S for the entire field are of interest. This is demonstrated in Figure 1 (A-2, B-2 and A-3, B-3), which depicts the profiles of the expectation (Equation [8]) and of the variance (Equation [9]) in terms of coefficient of variation (CV). It can be seen that the agreement between the two models ("precise" numerical and piston-type approximation) is quite good for the two first central statistical moments.

It is quite interesting to note in Figure 1 the dissimilarities between the shapes of $E(S)$ profiles as compared with the deterministic S profiles. This suggests that the expected value of water flow variables in heterogeneous fields of large variability does not necessarily look like the Richard's equation (Equation [1]) for a homogeneous equivalent field. This equation, which describes satisfactorily the physics of flow for a uniform soil column, fails to give the same results when it applies to heterogeneous fields. On the other hand, if the soil is of lesser variability, as in the case of Bet Dagan soil (see Bresler and Dagan, 1983a), the shapes of the expectation profiles of S look similar to deterministic profiles, and approximate and numerical profiles are not as close as in Figure 1.

III. Solute Transport

The most common approach to calculate solute concentration as a function of time and space is to model mass balance by a partial differential equation and, subsequently, to solve the equation for various initial and boundary conditions. The traditional numerical modeling approach is to regard the field as a homogeneous unit and to apply to it numerical solutions of the governing dispersion-convection equation for saturated-unsaturated soil valid for laboratory columns (or presuming that the field is a homogeneous porous medium). However, for spatially variable fields that, unlike small laboratory soil columns, display variation in their hydraulic properties (e.g., in K_s), the use of the traditional approach may not be useful. Instead, the actual heterogeneous field can be regarded approximately as a collection of vertical homogeneous columns differing in hydraulic properties, and, as a result, solute transport will differ from profile to profile depending on the local properties. In practice, one is neither interested nor able to measure the solute concentration at each point of the field, but the representation for the entire field is of interest. Regarding the hydraulic properties (e.g., K_s) as random, water content (θ), flow velocity ($V = q/\theta$), and, therefore, solute concentration are also random variables characterized by their PDF rather than by deterministic values. Hence, a rational approach for modeling solute transport in particularly variable soils is the stochastic one. This is the approach described in this chapter, and the statement of solute transport model on a field scale is quite different from the traditional one described in Bresler et al. (1982).

A. Mathematical Statement of Transport Problem

The macroscopic mass balance equation governing noninteracting solute dispersion in one-dimensional flow through a homogeneous column may be written as

$$\frac{\partial C}{\partial t} + V\frac{\partial C}{\partial z} = D_p\frac{\partial^2 C}{\partial z^2} + \frac{1}{\theta}\frac{\partial}{\partial z}\left(\theta D_h\frac{\partial C}{\partial z}\right) \qquad [40]$$

where C is the solute concentration, θ is the volumetric water content, z is the vertical coordinate positive downward, t is the time, D_h is the mechanical dispersion coefficient, and D_p is the effective molecular diffusivity for a static fluid in the soil system. A rough approximation for D_p is $(2/3)D_0$, where D_0 is the diffusivity in the solution phase. It will be assumed that $\theta = \theta(z, t)$ and the pore velocity, $V(z, t) = q/\theta$ (where q is the specific discharge), are given solutions of the flow equation (see Section II) independent of C. The concentration C is made dimensionless with respect to the difference between the initial concentration in the soil and that of the water applied to the soil surface. Hence, without loss of generality, C varies between zero and unity. For flow through saturated media ($\theta = \theta_s =$ constant), it has been shown that under quite general conditions (Saffman, 1959, 1960)

$$D_h = \lambda V \qquad [41]$$

in which λ is the dispersivity for longitudinal dispersion. Equations [40] and [41] are generally accepted, although with some reservations as valid for unsaturated flows (Bresler, 1973; Segol, 1977), with values of λ that vary in the range of 0.1–3 cm for most soils (Bresler and Laufer, 1974; Biggar and Nielsen, 1976).

Equation [40] refers to soil lying in the horizontal x,y plane and beneath $z = 0$ with z positive downward. Water flow takes place due to application of recharge q_0 on the surface, at the soil moisture content θ, and at the pore water velocity V. It is assumed that in spite of heterogeneity and eventual nonuniform application, flow is vertical. This assumption is bound to be quite accurate for the upper layer of thickness on the order of 1 to a few meters, which is considered here. Furthermore, Russo and Bresler (1981b) showed that the length scale characterizing the variation of hydraulic properties in the x,y plane is much larger than the depth (in the Bet Dagan field, at least) and, henceforth, lateral flows are negligibly small compared with the vertical flow. This assumption simplifies the problem considerably, and in the solute motion equation (Equation [40]), only derivatives with respect to t and z are maintained.

B. Stochastic Transport in Steady Gravitational Flow

The simplest solution of Equation [40] is obtained for uniform flow (i.e., $V =$ constant and $\theta =$ constant) and for initial and boundary conditions: $C = 1$ at $z = 0$, $C = 0$ at $t = 0$, $z > 0$, in a column extending along $z > 0$. Correspondingly, the next simplifying assumption is that the water flow is steady, that is, the vertical specific discharge q does not depend on z or t (q is constant in each profile, but varies in the x,y plane). This can be considered acceptable after the short time needed for the wetting front to move away from the soil surface has elapsed, and if the recharge q_0 does not change with time in the period of interest.

Under these conditions, the dispersion-convection equation (Equation [40]) becomes

$$\frac{\partial C}{\partial t} + V\frac{\partial C}{\partial z} = D\frac{\partial^2 C}{\partial z^2} \qquad [42]$$

with V and D independent of z.

Under these assumptions and for boundary and initial conditions,

$$\begin{aligned} C &= 1 \qquad z = 0, \quad t > 0 \\ C &= 0 \qquad z > 0, \quad t = 0 \end{aligned} \qquad [43]$$

the concentration profile of a noninteracting solute is described by the solution of Equation [42] subject to Equation [43] as follows:

$$C = (0.5)\left(1 - \operatorname{erf}\frac{z - Vt}{2(Dt)^{1/2}}\right) \qquad [44]$$

where C is the dimensionless concentration and $D = D_p + D_h$. The approximation given by Equation [44] is valid for $t \gg D/V^2$. Furthermore, D_p is generally negligible compared with D_h.

While for a homogeneous column V and D are constant under the above assumption, in a heterogeneous field they vary with x and y. Because of uncertainty, this variation is viewed as random and the actual field is regarded as a realization of an ensemble of fields with the same probability distribution functions (PDF) of the variables of interest. Furthermore, it is assumed that these PDF are statistically stationary, and under appropriate ergodic assumptions the ensemble averaging and space averaging in the given realization over x,y plane are equivalent.

Returning now to Equation [44], C is a random function of the random variables V and D (because of the random nature of V and θ), and it depends deterministically upon z and t. Hence, the value of C cannot be predicted deterministically, as it depends on the random variables V and θ. The main aim here, therefore, is to evaluate $f(z,t;C)$, the PDF of concentration for fixed z and t or the related cumulative probability:

$$P(z,t;C) = \int_0^A f(z,t;C)\,dC \qquad [45]$$

where P is the probability that $C \leqslant A$ at depth z and time t.

It should be remembered that under the ensemble and space equivalence, P represents the ratio between the area of the field whose solute concentration is less than A, and the total area, at given z and t.

With the aid of (f or P), one can evaluate a few moments of interest. The first moment, the average, is given by

$$E[C] = \int_0^1 Cf(C)\,dC = \int_0^1 C\,dP = CP\Big|_0^1 - \int_0^1 P\,dC = 1 - \int_0^1 P\,dC$$

Hence, the expectation

$$E[C(z,t)] = \int_0^1 Cf(z,t;C)\,dC = 1 - \int_0^1 P(z,t;C)\,dC \qquad [46]$$

is the average concentration of the field at depth z at time t. If the field is regarded as a fictitious equivalent homogeneous column, then it is reasonable to regard $E[C(z,t)]$ as the concentration in the equivalent field. The variation of C over the horizontal plane is characterized by the variance, the second central moment which is obtained as follows:

$$\begin{aligned}\sigma_c^2 &= \int_0^1 (C - C)^2 f(C)\,dC = \int_0^1 (C - \bar{C})^2\,dP \\ &= (\bar{C} - C)^2 P\Big|_0^1 - 2\int_0^1 (C - \bar{C}) P\,dC \\ &= (1 - \bar{C})^2 - 2\int_0^1 CP\,dC + 2\bar{C}\int_0^1 P\,dC\end{aligned}$$

Substituting $E[C]$ from Equation [46] for $\bar{C}$, yields

$$\begin{aligned}\sigma_c^2 &= (1 - \bar{C})^2 + 2\bar{C}(1 - \bar{C}) - 2\int_0^1 CP\,dC \\ &= 1 - \bar{C}^2 - 2\int_0^1 CP\,dC\end{aligned}$$

Thus, the second moment, the variance of $C(z,t)$, is calculated with the aid of $P(z,t;C)$ from

$$\begin{aligned}\sigma_c^2(z,t) &= \int_0^1 (C - \bar{C})^2 f(z,t;C)\,dC \\ &= 1 - \bar{C}^2 - 2\int_0^1 CP(z,t;C)\,dC\end{aligned} \qquad [47]$$

A measure of the symmetry of the distribution of C about its average is given by the skewness,

$$\begin{aligned}\gamma_c(z,t) &= \int_0^1 (C - \bar{C})^3 f(z,t;C)\,dC \\ &= (1 - \bar{C})^2 - 2\int_0^1 (C - \bar{C})^2 P(z,t;C)\,dC\end{aligned} \qquad [48]$$

The randomness of C satisfying Equation [44] stems from that of V and of D (D being in turn a function of V and θ). Hence, the moments of C have to be derived with the aid of the PDF of these two variables (because C is a function of V and θ).

The randomness of C is associated with the boundary and initial conditions in addition to its dependence on the variability of the soil variables. In the case of steady gravitational flow, which is the practical case of leaching applied on the soil surface, the rate q_0 (Equation [3]) is also a random variable denoted here after by R and characterized by a PDF $f(R)$. A simple realistic approximation to the PDF of the random variation of the recharge density R with x,y is the uniform (rectangular) PDF:

$$f(R) = 0 \quad \text{for} \quad R < \bar{R} - d_R \quad \text{and} \quad R > \bar{R} + d_R$$
$$f(R) = \frac{1}{2d_R} \quad \text{for} \quad \bar{R} - d_R < R < \bar{R} + d_R \qquad [49]$$

defined by the two parameters $\bar{R}$ and d_R so that the PDF of R is constant in the interval of width $2d_R$ around the average $\bar{R}$. In the case of rainfall over the field, d_R is very small, but for many irrigation methods or rainfall over orchards, $d_R/\bar{R}$ may be quite large.

1. Concentration Distribution for Piston Displacement: Closed-Form Solution

a. Theoretical Model for Average Concentration Distribution

In cases where dispersion may be insignificant in problems of solute displacement, an approximation of practical value (Reiniger and Bolt, 1972) for solute transport can be described by neglecting the diffusion and dispersion terms in Equation [40]. The convection equation without diffusion-dispersion can then be solved for any $q = V\theta$ in a closed form. Hence, C is constant for

$$\frac{dz}{dt} = \frac{q(z,t)}{\theta(z,t)} \qquad [50]$$

and for steady uniform flow, C is propagated along fronts of the equation

$$z = (q/\theta)t + \text{constant} = Vt + \text{constant} \qquad [51]$$

For the boundary and initial conditions leading to Equation [44], the value of C from Equation [51] is given by the simple expression

$$C(z,t) = H(Vt - z) \qquad [52]$$

where H is the Heaviside step function (i.e., $H(x) = 0$ for $x < 0$, $H(x) = 1$ for $x > 0$). Equation [52] is also the limit case of Equation [44], for $D \to 0$ representing a sharp front, separating the zone with $C = 1$ from that with $C = 0$, which moves downward at constant velocity V. In Equation [44], dispersion "smears" this front over a region of length (L).

Equation [40] can now be solved for given θ and V in terms of z and t at any given point of coordinates x and y in the field. Equivalently, in the statistical approach, V and θ are functions of the random parameters characterizing the field (e.g., K_s, η, β, θ_s, θ_r, ψ_w) and random parameters characterizing the flow (e.g., $q_0 = R$, t_i, θ_n). (To simplify matters, K_s has been singled out as the only stochastic parameter representing the field spatial variability, and R is the only one representing the flow at the boundary.) This will provide the dependence of the concentration profile on the coordinates x and y, which is random and can be expressed in terms of the frequency function (PDF) $f(K_s, R, \ldots,)$. Hence, one cannot answer the question about the exact magnitude of C at x, y, z, t in a spatially variable field, but only about the probability of C taking values between A and $A + dA$ at a depth z and time t. To derive explicitly a solution for the concentration probability distribution in a spatially variable field for steady gravitational flow and piston displacement, a steady recharge R with PDF (Equation [49]) is assumed. If irrigation or recharge R is applied for a sufficiently long time, a steady flow is achieved in each profile such that the flow is gravitational with constants θ and V. At any point in the field, one of two situations will occur: (1) if $R \geqslant K_s$, then in that portion of the field, ponding takes place, the flow is saturated, and with neglect of additional head on the surface (i.e., for unit gradient), Darcy's law yields

$$V = K_s/\theta_s; \qquad [53]$$

(2) for $R < K_s$, the soil is unsaturated and the velocity is given by

$$V = K(\theta)/\theta \quad \text{and} \quad K(\theta) = R \qquad [54]$$

Substituting Equation [14] into Equation [53], and Equations [6] and [7] (after neglecting θ_r, which is generally $< .05\theta_s$) into Equation [54] and inserting $Y = \ln \delta$ (Equation [12]) in Equation [14], yields for Equations [53] and [54]

$$V = \frac{K_s^* \exp(2Y)}{\theta_s} \quad \text{for } [R > K_s^* \exp(2Y)] \qquad [55]$$

$$V = \frac{R \exp(2\omega Y)}{\theta_s (R/K_s^*)^{\omega}} \quad \text{for } [R < K_s^* \exp(2Y)] \qquad [56]$$

where $\omega = \beta/\eta$. It can be shown that since $\theta_r \ll \theta_s$, dropping θ_r from Equation [7] would have a negligible effect upon the resulting V in Equation [55]. The two expressions, Equations [56] and [55], incorporate the entire kinematical information needed in order to solve Equation [40] for this simple flow configuration. Note that in Equations [55] and [56], V depends on the deterministic parameters K_s^* and θ_s and on the random variables R and Y, the PDF of which may be approximated by Equations [49] and [13]. The cumulative probability distribution of V can be computed in a general way from

$$P(V) = \int dY \int f(R, Y)\, dR \qquad [57]$$

where $f(R, Y)$ is the joint probability density function of R, Y and the integration region in the double integral depends on V and Y through Equations [56] and [55]. Because the recharge R and K_s are independent variables, then

$$f(R, Y) = f(R)f(Y) \qquad [58]$$

with $f(Y)$ and $f(R)$ given by Equation [13] and [49], respectively. The cumulative probability $P(V)$ can be calculated in an analytically closed-form solution of Equation [57] when $f(R, Y)$ is substituted from Equation [58].

In the simplified piston displacement model, the concentration (C) depends on the flow velocity (V) through Equation [52], from which C can be zero or unity. Hence, for given z and t, if $V > z/t$ then $C = 1$, but if $V < z/t$ then $C = 0$. The PDF of C is therefore represented by two Dirac distributions at $C = 0$ and $C = 1$ (see Dagan and Bresler, 1979, for details), so that the area of the spike in $C = 0$ is

$$P(\varepsilon) = P(V \leqslant z/t) \qquad [59]$$

Conversely, the area of the spike at $C = 1$ is

$$1 - P(V \leqslant z/t) \qquad [60]$$

where $\varepsilon \to 0$. Substituting $V = z/t$ in Equations [54], [55], and [56], the cumulative probability (Equation [59] or [60]) is obtained. Since $P(z/t)$ gives the ratio for a given t and at a depth z between the area of a field for which $C = 0$ and the total area, the average concentration ($\bar{C}$) over the field is given by

$$\bar{C}(z, t) = 1 - P(z/t) = 1 - P(V) \qquad [61]$$

Physically, $\bar{C}$ is the ratio between the area of the field for which $C = 1$ and the total area of the field. This is because in this simplified piston displacement model, the concentration front separating the zone of $C = 1$ from the zone of $C = 0$ is a surface propagating downward at velocity $V = z/t$, which changes from profile to profile due to the variability of K_s (or Y). For example, at the soil surface ($z = 0$), the cumulative probability $P(V = z/t = 0) = 0$ and therefore $\bar{C} = 1$, but on the other hand, at large z and small t $P(z/t) = 1$ and therefore $\bar{C} = 0$. At intermediate z values of C ($0 \leqslant C \leqslant 1$) can be calculated in closed form as detailed in the next section.

b. Closed-Form Calculations of Average Field Concentration $\bar{C}(z, t)$

To calculate $\bar{C}$ from Equation [61], the cumulative probability $P(V = z/t)$ has to be computed by Equation [57] (with Equation [58] substituted for $f(R, Y)$) with the entire domain of integration $-\infty < Y < \infty$, $\bar{R} - d_R < R < \bar{R} + d_R$. The complication in calculating $\bar{C}$ in a closed analytical form, and therefore the lengthy equation of $\bar{C}$, arises from the existence of ponding over part of the field area. Consequently, there is a need to split the calculations into two different expressions with a border between ponding and nonponding at $R = K_s^* \exp(2Y)$, that is, at

$$Y = 0.5\ln(R/K_s)$$

The end points are at

$$Y = 0.5\ln\frac{\bar{R} - d_R}{K_s^*} \quad \text{and} \quad Y = 0.5\ln\frac{\bar{R} + d_R}{K_s^*}.$$

For the range of recharge or irrigation rate $\bar{R} - d_R \leqslant R \leqslant \bar{R} + d_R$, there are, in general, three cases: (1) ponding because $V < (\bar{R} - d_R)/\theta_s$ so that (Equation [55])

$$Y \leqslant Y_2 = 0.5\ln(V\theta_s K_s^*) \tag{62}$$

(2) Nonponding because $V > (\bar{R} + d_R)/\theta_s$ so that (Equation [56])

$$Y \leqslant Y_3 = (1/2\omega)\ln\frac{V\theta_s(R/K_s^*)^\omega}{R} \tag{63}$$

with

$$Y_b = (1/2\omega)\ln[V\theta_s(\bar{R} - d_R)^{-1}(K_s^*)^{-\omega} \tag{64}$$

and if $Y_a > Y_b$ then

$$Y_a = (1/2\omega)\ln[V\theta_s(\bar{R} + d_R)^{\omega-1}(K_s^*)^{-\omega}] \tag{65}$$

(3) Flow velocity is within the whole range of R, that is, $(\bar{R} - d_R)/\theta_s < V < (\bar{R} + d_R)/\theta_s$.

To derive expressions for $P(V)$ in each of these three cases, normal PDF $f(Y)$ from Equation [13] and uniform PDF $f(R)$ (Equation [49]) are substituted to Equation [58] and then to Equation [57] and are integrated over the relevant domain. For this integration, it is convenient to use the following auxiliary formulas:

$$\int_{-\infty}^{A} e^{\alpha Y} f(Y)\,dY = 0.5\exp\left(\alpha m_Y + \frac{\alpha^2\sigma_Y^2}{2}\right)\left(1 + \operatorname{erf}\frac{A - m_Y - \alpha\sigma_Y^2}{\sqrt{2}\sigma_Y}\right) \tag{66}$$

$$\int_{B}^{A} e^{\alpha Y} f(Y)\,dY = 0.5\exp\left(\alpha m_Y + \frac{\alpha^2\sigma_Y^2}{2}\right) \times \left(\operatorname{erf}\frac{A - m_Y - \alpha\sigma_Y^2}{\sqrt{2}\sigma_Y} - \operatorname{erf}\frac{B - m_Y - \alpha\sigma_Y^2}{\sqrt{2}\sigma_Y}\right) \tag{67}$$

This yields for the above-mentioned three cases the following expressions:
(1) For $V < (\bar{R} - d_R)/\theta_s$,

$$P(V) = \int_{\bar{R}-d_R}^{\bar{R}+d_R} dR \int_{-\infty}^{Y_2} dY\, f(R) f(Y)$$

which gives, after integration, using Equation [66] with $\alpha = 0$

$$P(V) = (0.5)\left[1 + \operatorname{erf}\frac{(0.5)\ln\dfrac{V\theta_s}{K_s^*} - m_Y}{(2)^{1/2}\sigma_Y}\right] \tag{68}$$

(2) For $V > (R + d_R)/\theta_s$,

$$P(V) = \int_{\bar{R}-d_R}^{\bar{R}+d_R} dR \int_{-\infty}^{Y_3} dY\, f(R)f(Y)$$

$$= \int_{R-d_R}^{R+d_R} dR \int_{-\infty}^{Y_a} dY\, f(R)f(Y) + \int_{Y_a}^{Y_b} dY \int_{R-d_R}^{R(Y_3)} dR\, f(Y)f(R)$$

which gives, after integration, using Equation [66] for the first term on the right-hand side and Equation [67] for the second term,

$$\begin{aligned} P(V) = {} & 0.5\left[1 + \operatorname{erf}\frac{Y_a - m_Y}{(2)^{1/2}\sigma_Y}\right] - \left[\frac{\bar{R} - d_R}{4d_R}\right]\left[\operatorname{erf}\frac{Y_b - m_Y}{(2)^{1/2}\sigma_Y} - \operatorname{erf}\frac{Y_a - m_Y}{(2)^{1/2}\sigma_Y}\right] \\ & + 0.25d_R(V\theta_s/K_s^{*\omega})^{1/(1-\omega)}\exp[(-2\omega m_Y/(1-\omega)] \\ & + (2\omega^2\sigma_Y^2/(1-\omega)^2] \\ & \times\left[\operatorname{erf}\frac{Y_b - m_Y + (2\omega\sigma_Y^2)/(1-\omega)}{(2)^{1/2}\sigma_Y} - \operatorname{erf}\frac{Y_a - m_Y + (2\omega\sigma_Y^2)/(1-\omega)}{(2)^{1/2}\sigma_Y}\right] \end{aligned} \quad [69]$$

where Y_b and Y_a are given explicitly by Equations [64] and [65], respectively.

(3) For $(\bar{R} - d_R)/\theta_s < V < (\bar{R} + d_R)/\theta_s$,

$$P(V) = \int_{\bar{R}-d_R}^{\bar{R}+d_R} dR \int_{-\infty}^{Y_2} dY\, f(R)f(Y) + \int_{Y_2}^{Y_b} dY \int_{R-d_R}^{R(Y)} dR\, f(Y)f(R)$$

which gives, after integration, using Equations [66] and [67],

$$\begin{aligned} P(V) = {} & 0.5\left[1 + \operatorname{erf}\frac{Y_2 - m_Y}{(2)^{1/2}\sigma_Y}\right] - \frac{\bar{R} - d_R}{4d_R}\left[\operatorname{erf}\frac{Y_b - m_Y}{(2)^{1/2}\sigma_Y} - \operatorname{erf}\frac{Y_2 - m_Y}{(2)^{1/2}\sigma_Y}\right] \\ & + \frac{1}{4d_R}\left(\frac{V\theta_s}{K_s^{*\omega}}\right)^{1/(1-\omega)}[\exp(-2\omega m_Y/(1-\omega) + (2\omega^2\sigma_Y^2/(1-\omega)^2] \\ & \times\left[\operatorname{erf}\frac{Y_b - m_Y + (2\omega\sigma_Y^2)/(1-\omega)}{(2)^{1/2}\sigma_Y} - \operatorname{erf}\frac{Y_2 - m_Y + 2\omega\sigma_Y^2/(1-\omega)}{(2)^{1/2}\sigma_Y}\right] \end{aligned} \quad [70]$$

where Y_2 is given by Equation [62].

In the particular simple case of uniform water application or recharge, that is, $d_R = 0$ (or $S_R = d_R/\bar{R} = 0$),

$$\bar{C} = 0.5\left[1 - \operatorname{erf}\frac{0.5\ln\xi - m_Y}{(2)^{1/2}\sigma_Y}\right] \quad \text{for } (\xi < r) \quad [71]$$

$$\bar{C} = 0.5\left[1 + \operatorname{erf}\frac{(0.5\omega)\ln(\xi r^{\omega-1}) - m_Y}{(2)^{1/2}\sigma_Y}\right] \quad \text{for } (\xi > r) \quad [72]$$

where

$$\xi = z\theta_s/tK_s^*, \qquad r = \bar{R}/K_s^*, \quad \text{and} \quad S_R = d_R/\bar{R} \quad [73]$$

are convenient dimensionless forms. Note that the definitions in Equation [73] enable one to compute from Equations [61], [68], [69], and [70] the value of $\bar{C}$ in terms of the dimensionless variables ξ, r, and S_R, in addition to m_Y, σ_Y, and ω.

c. Illustration of Results

To calculate average concentration profiles $\bar{C}(z,t)$, formulated by Equations [61]–[73], for a particular field, the values of the parameters K_s^*, m_Y, σ_Y, and ω must be known. Then, evaluation of $\bar{C}(z,t)$ is straightforward for any combination of $r = \bar{R}/K_s^*$ and $S_R = d_R/\bar{R}$, characterizing the average and variance of the rate of application and of the soil parameters characterizing the field. In Figure 2, examples of model calculations of the average concentration profiles are given for Panoche soil series studied extensively by Nielsen et al. (1973) and Warrick et al. (1977), with a value of $1/\omega = 7.2$, adapted from Bresler et al. (1978). The other values of the parameters used to calculate the results of Figure 2 were for m_Y and σ_Y those of Equation [39], and the value of $K_s^* = 0.22$, which was calculated from the known estimation of $E(K_s) = 0.93$ and the definitions in Equations [12], [14], and [15], was as follows: $E(K_s) = K_s^* E[\exp(2Y)] = K_s^* \int \exp(2Y) f(Y)\,dY$ and with $f(Y)$ normal, $K_s^* = E(K_s)[\exp(-2m_Y - 2\sigma_Y^2) = 0.93 \exp(-1.46) = 0.22$ cm/hr. A few sets of curves are represented in Figure 2 for $S_R = 0$ (i.e., deterministic uniform water application on the soil surface) and $S_R = 1$ (i.e., $d_R = \bar{R}$, the largest CV of R) for three values of $r = \bar{R}/K_s^*$ (i.e., 0.2, 0.6, and 1.0).

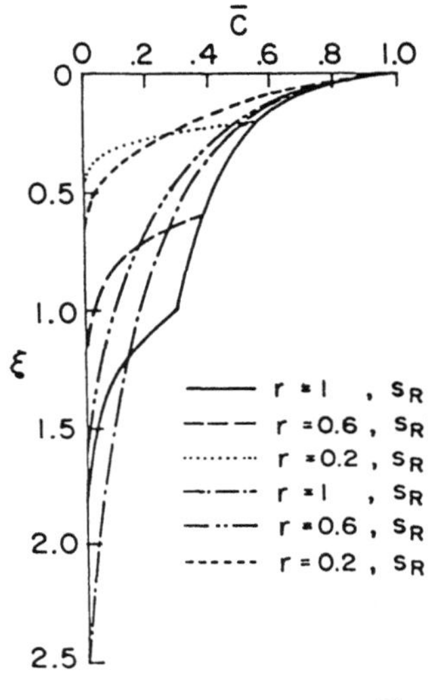

Figure 2. Computed average concentration profile ($\bar{C}$) as a function of $\xi = z\theta/tK_s^*$ for three values of the dimensionless application rate ($r = \bar{R}/K_s$) for deterministic ($S_R = 0$) and stochastic ($S_R = 1$) rates.

Consider first the simplest case of $S_R = 0$ (i.e., deterministic uniform recharge on the soil surface). In this case, the concentration profile for the zone extending from the soil surface to $\xi = r$ (i.e., $\xi \leqslant r$) is described by Equation [71], regardless of r. This is demonstrated by the curves lying above the breaking points of Figure 2. In this zone, which lies between $\xi = 0$ and $\xi = r$, solute transport is controlled by the saturated flow beneath the ponded area. For the zone in which $\xi > r$, the $\bar{C}(\xi)$ profile is described by Equation [72], which pertains here to the unsaturated flow. Hence, the three $S_R = 0$ curves of Figure 2 differ mainly in the location of the breaking point of change of shape of the profile at $\xi = r$ and $\bar{C}(\xi)$ profiles for other values of r can be obtained easily by translation. For cases where $S_R > 0$, which are demonstrated by the curves of Figure 2 corresponding to $S_R = 1$, the $\bar{C}(\xi)$ profiles are calculated by Equations [68] through [70]. Here the curves are smooth as $\bar{C}$, at a given depth, is influenced by front translation either in the saturated or in the unsaturated zone because of the simultaneous but independent, variation of K_s and R.

Figure 2 shows that for a given S_R, the solute spreading over the entire field is greater for large values of r than for small values. This is because a larger portion of the field is ponded for large r and then the saturated flow zone plays a dominant role. In this zone, the front velocity varies in a wide range because of relatively high K_s variation. In contrast, in the unsaturated part of the field, the variation of V depends on θ (Equation [54]), which varies less widely than K_s because of the algebraic relationship in Equations [6] and [7]. It is emphasized that for smaller values of $1/\omega$, or if ω would also have been considered as a random variable, solute spread should become larger than that demonstrated in Figure 2.

2. Concentration Distribution for Dispersive Transport: Numerical Quadrature Solution

a. Derivation of PDF for Solute Concentration

To relate the variables V and D in Equation [44] to the spatially variable soil properties and the rate of application boundary conditions, the stochastic steady flow velocity (V) is determined from Equations [6], [7], [12], [13], [14], [49], [53], and [54]. The additional factor considered in this section is the dispersion coefficient (D) in Equations [40] through [44]. Taking Equation [41] as the representative expression for $D(V)$, the soil parameter, λ, i.e., the dispersivity, fits lognormal distribution (Biggar and Nielsen, (1976). Hence, if we define μ as

$$\mu = \ln \lambda \qquad [74]$$

then $f(\mu)$ is given by Equation [13] after replacing Y by μ with the statistical parameters σ_μ and m_μ.

Now C is a function of V and λ (Equations [41] and [44]), which, in turn, are functions of Y, R, and μ (Equations [6], [7], [12], [13], [14], [49], [53], [54], and [74]). The PDF of C can then be written in a general manner

$$f(C)\,dC = f(Y, R, \mu)\,dY\,dR\,d\mu \qquad [75]$$

Furthermore, it is assumed here, for the sake of simplicity, that Y, R, and μ are independent random variables. This is obvious for R and the two others, but it is less plausible for Y and μ, which are both soil properties. Hence, Equation [75] becomes

$$f(C)\,dC = f(Y)f(R)f(\mu)\,dY\,dR\,d\mu \qquad [76]$$

with $f(Y)$ (Equation [13]), $f(R)$ (Equation [49]), and $f(\mu)$ similar to Equation [13].

Summarizing the procedure in principle, the input parameters of the problem are K_s^*, D_p, θ_r, θ_s, β, m_Y, σ_Y, $\bar{R}$, d_R, m_μ, and σ_μ. They determine entirely the random variable V by Equations [6], [7], [12], [13], [14], [49], [53], and [54] and, therefore, $f(C)$ (Equation [76]) is defined by them as well. The detailed computational scheme is discussed next.

b. Computational Procedure

To simplify the computations, only two of the three variables Y, R, and μ are considered as random and the third one as deterministic. The basic function to be calculated is the cumulative probability of C (Equation [45]). In the first case in which Y and μ are random while R is deterministic and equal to $\bar{R}$, $P(C)$ is given by the combination of Equations [45], [13], and [76], which yields

$$P(z, t; C) = \frac{1}{(2\pi)^{1/2}\sigma_\mu}\int_{-\infty}^{\infty} \exp\left[-\frac{(\mu - m_\mu)^2}{2\sigma_\mu^2}\right] d\mu \int_{-\infty}^{Y} \frac{1}{(2\pi)^{1/2}\sigma_Y} \times \exp\left[-\frac{(\alpha - m_Y)^2}{2\sigma_Y^2}\right] d\alpha \qquad [77]$$

where $Y = Y(z, t; C, \mu, \bar{R})$.

By integrating once over α, Equation [77] becomes

$$P(z, t; C) = \frac{1}{2(2\pi)^{1/2}\sigma_\mu}\int_{-\infty}^{\infty} \left\{\exp\left[-\frac{(\mu - m_\mu)^2}{2\sigma_\mu^2}\right] \cdot \left[1 + \operatorname{erf}\frac{Y(z, t; C, \mu, \bar{R}) - m_Y}{\sqrt{2}\sigma_Y}\right]\right\} d\mu \qquad [78]$$

so that P is determined by a quadrature that has to be carried out numerically. By the same token, for Y, R (random), and $\lambda = \bar{\lambda}$ (deterministic), one has by Equations [45], [13], [49], and [76]

$$P(z, t; C) = \int_{R-d_R}^{R+d_R} f(R)\,dR \int_{-\infty}^{Y} f(Y)\,dY = \frac{1}{4d_R}\int_{\bar{R}-d_R}^{\bar{R}+d_R} \left[1 + \operatorname{erf}\frac{Y(z, t; C, \bar{\lambda}, R) - m_Y}{\sqrt{2}\sigma_Y}\right] dR \qquad [79]$$

Again P is obtained by a numerical quadrature with negligible truncation error.

The function $Y(z, t; C, \lambda, R)$, which appears in Equations [78] and [79], is obtained by inverting Equation [44] after substituting Equation [41] for D, as follows:

$$\frac{z - V(R, Y)t}{2(\lambda V(R, Y)t + D_p t)^{1/2}} = \phi(C) \qquad [80]$$

where

$$\phi(C) = \mathrm{erf}^{-1}(1 - 2C) \qquad [81]$$

From Equation [80], which is a quadratic equation in V, one obtains

$$V = [z + 2\phi^2\lambda - 2\phi(\lambda z + \phi^2\lambda^2 + D_p t)^{1/2}]/t \qquad [82]$$

This can be rewritten also with the aid of the dimensionless variable $\xi = z\theta_s/tK_s^*$ replacing t (Equation [73]) as follows:

$$V = \frac{K_s^*\xi}{\theta_s} + \frac{K_s^*\xi}{z\theta_s}\left[2\phi^2\lambda - 2\phi\left(\lambda z + \phi^2\lambda^2 + \frac{D_p z\theta_s}{\xi K_s^*}\right)^{1/2}\right]$$

With these preparatory steps, the computation of $P(z, t; C)$ is carried out as follows:

1. Given values of z, t, C (i.e., ϕ of Equation [81]), D_p, $\lambda = \exp(\mu)$, and R are substituted in Equation [82] to obtain V.
2. To find out which one of the two relationships Equation [53] or [54] holds, V is compared with R/θ_s, which is the value separating unsaturated flow (Equation [54]) from ponding (Equation [53]). If $V < R/\theta_s$, ponding takes place and Equation [53] is valid. Then, by Equations [14] and [12]

$$V = \frac{K_s}{\theta_s} = \frac{K_s^* \exp(2Y)}{\theta_s} \qquad [83]$$

or

$$Y = (1/2)\ln(V\theta_s/K_s^*) \qquad [84]$$

If $V > R/\theta_s$, the flow is unsaturated, Equation [54] is valid, and by Equations [6], [7], [14], and [12],

$$V = \frac{R}{\theta} = R/[\theta_r + (R/K_s^*)^{\omega}(\theta_s - \theta_r)\exp(-2\omega Y)] \qquad [85]$$

or

$$Y = \frac{1}{2\omega}\ln\frac{(R/V) - \theta_r}{(R/K_s^*)^{\omega}(\theta_s - \theta_r)} \qquad [86]$$

Hence, for given θ_r, θ_s, K_s^*, and ω, $Y(z, t; C, R, \mu)$ is determined from Equation [84] or [86].

3. This value of Y is substituted in Equation [78] with $R = \bar{R}$ or in Equation [79] with $\lambda = \bar{\lambda}$, and P is determined by numerical integration by augmenting the values of μ or R by small increments.
4. Once $P(z, t; C)$ has been determined numerically for a set of C values, various moments of C are found from Equations [46] and [47] by an additional integration over C, in Equation [78] or [79].

c. Illustration of Results

To illustrate the capability of the model $P(z, t; C)$, computations from Equations [78] and [79], which provide the most detailed information, are presented in Figure 3 for five values of z and at a given t. For these computations, values of m_Y and σ_Y are those of Equation [39] adopted from Warrick et al. (1977); and values of m_μ, σ_μ, and $\bar{\lambda}$ are calculated from Biggar and Nielsen (1976) by Bresler and Dagan (1981) as

$$m_\mu = 0.41, \qquad \sigma_\mu = 1.17, \qquad \text{and} \qquad \bar{\lambda} = 3 \text{ cm} \qquad [87]$$

The values of the other parameters are those of Panoche soil with $\theta_r \approx 0$, $\theta_s = 0.43$, $K_s^* = 0.22$ cm/hr and with $S_R = 0$.

The curves $P(z, t; C)$ in Figure 3 have physical interpretation. They represent the area of the field relative to the total area, which is at a concentration smaller than C at depth z and time t. The application of curves like those in Figure 3 is immediate if one wishes to know what proportion of the field at depth z is at a concentration larger than, say, $C = 0.5$ after, say, $t = 24$ hours of irrigation with water at a dimensionless concentration of $C = 1$.

After evaluating $P(z, t; C)$ the average concentration distribution $\bar{C}(z, t)$, the variance $\sigma_C^2(z, t)$, and the skewness $\gamma_C(z, t)$ can be computed from Equations [46], [47], and [48], respectively, by numerical evaluation of the integrals. Various combinations of the parameters $r = \bar{R}/d_R$, K_s^* and $S_R = d_R/\bar{R}$ can be

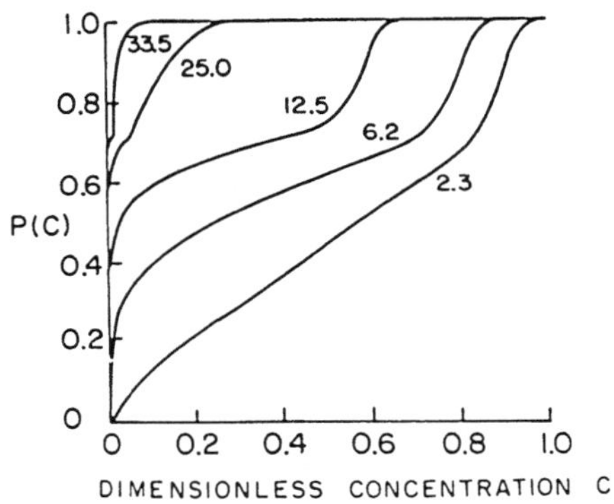

Figure 3. Cumulative probability $P(z, t; C)$ as a function of concentration C, for various z in cm (the numbers labeling the lines) and fixed $t = 24$ hr computed with $m_\mu = 0.41$, ($\lambda = 3$ cm), $\sigma_\mu = 1.17$, $D_p = 0.02$ cm^2/hr, $\sigma_Y = 1.16$, $r = 0.1$, and $S_R = 0$.

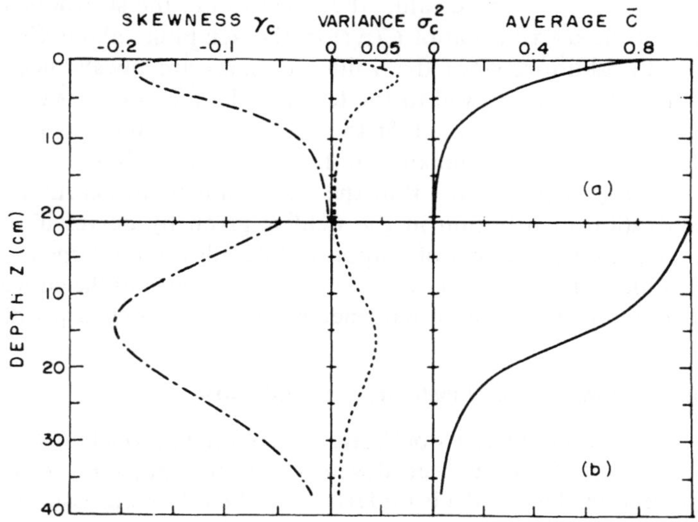

Figure 4. The first ($\bar{C}$), second (σ_C^2), and third (γ_C) moments of C as a function of soil depth (z) for fixed infiltration time (t). (a) Computations for Panoche soil with $t = 24$ hr, $m_Y = -0.62$, $\sigma_Y = 1.16$, $\bar{\lambda} = 3$ cm, $\sigma_\mu = 0$, $r = 0.2$, $S_R = 0$; (b) Bet Dagan soil with $\bar{\lambda} = 3$ cm, $r = 0.2$, $\sigma_Y = 0.4$, $m_Y = -0.1$, and $t = 5$ hr.

used. The three moments characterizing the function $f(z, t, C)$ (i.e., Equations [46], [47], and [48]) are illustrated in Figure 4 for two soils: Panoche, with high variability in K_s and low average K_s; and Bet Dagan, with low variability and high K_s ($m_Y = -0.1$, $\sigma_Y^2 = 0.16$).

The first moment (given by Equation [46]) represents the concentration at depth z and time t averaged over the entire field in the x,y plane. When comparing the $\bar{C}(z)$ profiles of Panoche (Figure 4a) and Bet Dagan (Figure 4b), it is immediately seen that the shape in Panoche soil is far from the sigmoid curve given by Equation [44], while the shape of $\bar{C}(z)$ for Bet Dagan resembles a sigmoid. This difference is related to the lesser variability of hydraulic conductivity of Bet Dagan soil.

An additional parameter of interest, illustrated in Figure 4, is the second central moment: the variance of the concentration $\sigma_C^2(z, t)$, being defined in Equation [47]. It characterizes the degree of variation of the concentration over the horizontal plane at a depth z and time t. It is seen that $\sigma_C = 0$ at the surface ($z = 0$), where by definition $C = 1$ and that $\sigma_C(z)$ also vanishes at larger values of z when $C \to 0$. Maximum values of $\sigma_C(z)$ are at $z = 2.5$ cm, $t = 24$ hr in Panoche soil and $z = 17$ cm, $t = 5$ hr in Bet Dagan soil. At the same depths, the average concentration is $\bar{C} = 0.5$, precisely as found in the absence of pore scale dispersion.

The third central moment, the skewness, being defined by Equation [48], is also illustrated in Figure 4. Being a measure of the symmetry of the distribution about its average value, the deviation of the skewness from zero indicates that the distribution of $C(z,t)$ in the x,y plane about $\bar{C}(z,t)$ is not symmetric. In addition, since the values of $\gamma_C(z)$ are always negative, the distribution of $C(z,t)$ is skewed to the left (i.e., the mean of C is to the left of its median, which in turn is left of the mode of C). Here, also, maximum skewness is associated with maximum deviation and γ_C is zero at $z = 0$ and vanishes for large depths. Note that the entire statistical information about the concentration distribution in the field is given by curves like those in Figures 3 and 4. If the number of graphs and details are too large for application, this model can be used to examine the dependence of the concentration distribution upon various input parameters for this particular problem.

C. Stochastic Transport in Nonsteady Vertical Flow

In Section I of this chapter, a stochastic approximate model of water flow in a spatially variable field has been developed with application to infiltration and redistribution. The model is underlain by the following assumptions: the flow is vertical, spatial variability occurs in the horizontal plane, variability is associated with the saturated conductivity, deterministic uniform recharge is applied on the soil surface during $t \leqslant t_i$ and the soil is with uniform $\theta = \theta_n$ prior to infiltration, and with a piston-type water content profile is assumed. In this part, the problem of inert solute transport under the same water flow conditions is modeled. Solute concentration prior to infiltration (C_n) is uniform, while water with uniform concentration $C_0 > C_n$ is applied at the surface during $0 < t < t_i$. The aim is to compute the expectation $E(C)$ and the variance σ_C^2 as a function of z and t under these conditions (these are quantities of interest in many applications to leaching- or pollution-related problems). For this purpose, a simplified model of salt transport and a more complicated and accurate computer model are developed to compute mean concentration and its variance in a spatially variable field.

1. Derivation of Approximate Solution of Salt Transport

For vertical one-dimensional flow of deterministic pore velocity $V(z,t)$, the equations of transport (Equations [40] and [43]) for an inert solute can be written as follows:

$$\frac{\partial C}{\partial t} + V\frac{\partial C}{\partial z} = \frac{\partial}{\partial z}\left(D\frac{\partial C}{\partial z}\right) \qquad [88]$$

$$C = 0 \qquad t = 0 \qquad z > 0 \qquad [89]$$

$$C = 1 \qquad 0 < t < t_i \qquad z = 0 \qquad [90]$$

$$\partial C/\partial z = 0 \qquad t_i < t \qquad z = 0 \qquad [91]$$

Adopting the approximate solution for the water flow developed in Section I, it follows that the actual moisture content profile is replaced by one of uniform

$\theta = \bar{\theta}$ or $\bar{S} = (\bar{\theta} - \theta_r)/(\theta_s - \theta_r)$ extending from the soil surface $z = 0$ to the equivalent front $z = L$, while ahead of the front ($z > L$) the moisture content has the initial θ_n. In the stage of infiltration with constant recharge q_0 applied on $z = 0$ and for redistribution as well, $\bar{V} = \bar{q}/\bar{\theta} = \bar{q}[\bar{S}(\theta_s - \theta_r) + \theta_r]$ is also constant in the profile and depends on t solely.

Hence, the equations satisfied by the concentration C, replacing Equation [88] under piston water flow conditions, become

$$\frac{\partial C}{\partial t} + \bar{V}(t)\frac{\partial C}{\partial z} = \frac{\partial}{\partial z}\left(D\frac{\partial C}{\partial z}\right) \qquad 0 < z < L \qquad [92]$$

To simplify the problem further, $\bar{V}$ is related to L by the equation

$$\bar{V}(t) = dL/dt \qquad [93]$$

which is exact in the gravitational, late stage of infiltration. Note that Equation [93] is quite accurate even at the beginning of the infiltration process. Another simplification is achieved by neglecting the molecular diffusivity term so that Equation [92] becomes

$$\frac{\partial C}{\partial t} + \frac{dL}{dt}\frac{\partial C}{\partial z} = \lambda\frac{dL}{dt}\frac{\partial^2 C}{\partial z^2}; \qquad 0 < z < L \qquad [94]$$

Carrying out the change of variables

$$\zeta = z - L(t); \qquad \tau = \int_0^t \frac{\lambda}{\lambda_{\max}}\frac{dL}{dt'}dt' \qquad [95]$$

gives $C(z, t) = C(\zeta, \tau)$. Substituting Equation [95] into Equation [94] yields

$$\frac{\partial C}{\partial \tau} = \lambda_{\max}\frac{\partial^2 C}{\partial \zeta^2} \qquad -L < \zeta < 0 \qquad [96]$$

while the initial and boundary conditions for Equations [89], [90], and [91] become, respectively,

$$C = 0 \qquad \tau = 0 \qquad \zeta > 0 \qquad [97]$$

$$C = 1 \qquad \zeta = -L(\tau) \qquad [98]$$

$$\frac{\partial C}{\partial \zeta} = 0 \qquad t > t_i \qquad \zeta = -L \qquad [99]$$

In the moving frame ζ (Equation [95]), the problem is therefore of heat conduction with conditions in Equation [98] or [99] along the moving boundary $\zeta = -L$. An additional boundary condition needed in order to solve Equation [96] is the one prevailing at the front $z = L$, $\zeta = 0$. In a numerical accurate solution of the flow and transport problem, the appropriate condition is $\partial C/\partial z = 0$ for $z = L_F(t)$. However, by the definition of L in Section I, the equivalent front is above the actual one, (i.e., $L < L_F$). To simplify matters, it is assumed that the condition $\partial C/\partial z = 0$ applies sufficiently far from $z = L$ to justify adopting for C the solution pertaining to a vertical column of indefinite length. Furthermore, except for exceedingly short times, L is sufficiently large to allow removing boundary conditions in Equations [98] and [99] from $\zeta = -L$ to $\zeta = -\infty$. Under these simplifying conditions, the solution of Equation [96] is given by (Carslaw and Jaeger, 1959):

$$C(\zeta, \tau) = \mathrm{erf}\left[-\frac{\zeta}{2(\lambda_{\max}\tau)^{1/2}}\right] \qquad 0 > \zeta > -L \qquad [100]$$

By the definition of τ and with the assumed variation of λ from $\lambda = 0$ to $\lambda = \lambda_{\max}$ (Bresler and Dagan, 1983b), one has, from Equation [95],

$$\tau = \frac{L^2}{26\lambda_{\max}} \qquad \text{for } \frac{L}{\lambda_{\max}} < 13 \qquad [101]$$

$$\tau = L - \frac{13}{2}\lambda_{\max} \qquad \text{for } \frac{L}{\lambda_{\max}} > 13 \qquad [102]$$

Hence, $C(\zeta, \tau)$ of Equation [100] is given by the following approximate expressions, after substitution of Equations [95], [101], and [102]

$$C(z, t; Y) = \mathrm{erf}\left[\frac{L(t; Y) - z}{(2/26)^{1/2}L(t; Y)}\right] \qquad L < 13\lambda_{\max} \qquad [103]$$

$$C(z, t; Y) = \mathrm{erf}\left[\frac{L(t; Y) - z}{2\lambda_{\max}^{1/2}(L - 6.5)^{1/2}}\right] \qquad L > 13\lambda_{\max} \qquad [104]$$

It is assumed, somewhat arbitrarily, that for $L > 13\lambda_{\max}$ the dispersivity becomes constant and equal to $\lambda_{\max}$.

The derivation of the expectation $E(C)$ and the variance σ_C^2 is straightforward by substituting Equations [103] and [104] (with values of $L(t; Y)$ determined explicitly during infiltration and redistribution in Section I)

$$E[C(z, t)] = \int_{-\infty}^{\infty} C(z, t; Y) f_Y(Y)\, dY \qquad [105]$$

and

$$\sigma_C^2(z, t) = \int_{-\infty}^{\infty} (C - E[C])^2 f_Y(Y)\, dY \qquad [106]$$

for f_Y normal as assumed to be a normal PDF defined by Equation [13]. As a matter of fact, one can also determine $f_C(z, t; C)$, the PDF of C, by using the relationships

$$f_C\, dC = f_Y\, dY \qquad P_c = \int_0^C f_c\, dC = \int_{-\infty}^{Y(z,t;C)} f_Y\, dY \qquad [107]$$

which, in terms of spatial distribution, permits one to evaluate the fractional area of the entire field for which C is smaller than a given number.

Note that the integration over $f_Y\,dY$ is also performed in Section I when computing the various statistical moments of S at different z and t (Equations [8] and [16]), and it is extremely simple to incorporate C in the same integration process.

2. Computations of Statistical Moments of C

With values of m_Y, σ_Y, and six deterministic parameters (K_s^*, θ_s, θ_r, β, η, and ψ_w, characterizing the field), the desired moments of $C(z,t)$ are computed for various combinations of the initial and boundary parameters ($R = q_0$, t_i, and θ_n, characterizing the flow).

The mean $E[C]$ and the variance σ_C^2 have been computed by two methods: (1) by using the numerical exact solution of the water flow problem and, subsequently, the numerical solution by finite differences as described by Bresler (1973); and (2) by employing the approximate model of water flow to calculate $L(t; Y)$ from Equation [22], which is substituted in Equation [103] or [104] to obtain $C(z,t; Y)$.

The computation procedure for the approximate model involves the following steps:

1. Set $i = 1$, define t, and calculate $L_i(t)$ from Equation [22]

$$L_i(t) = V_i(t)/[(\theta_s - \theta_r)(\bar{S} - S_n)]$$

2. Define and list several values of z.
3. Compare $L_i(t)$ with the input values of z. If $L_i(t) \geqslant z$, compute for each input value of z, $C_i(z,t) = C(z,t; Y_i)$ from Equations [103] and [104]. Otherwise, $C_i(z,t) = 0$.
4. Set $i = 2, 3, \ldots, N$ and repeat steps 1–3.
5. Once $C_i(z,t)$ has been determined for the whole set of N values, the three moments of C (averages, variances, and skewness) are computed from

$$E[C(z,t)] = \frac{1}{N}\sum_{i=1}^{N} C_i(z,t) \qquad [108a]$$

$$\sigma_C^2(z,t) = \frac{1}{N}\sum_{i=1}^{N} \{C_i(z,t) - E[C(z,t)]\}^2 \qquad [108b]$$

$$\gamma_C(z,t) = \frac{1}{N}\sum \{C_i(z,t) - E[C(z,t)]\}^3 \qquad [108c]$$

6. Change t and repeat Steps 1–5 above.

For the numerical model, finite difference equations to approximate the partial differential equation (Equation [88]), the initial conditions Equation [89], and the boundary conditions (Equation [91]) are first formulated and then solved numerically (for details, see Bresler, 1973). The boundary condition (Equation [90]) assigned at the soil surface ($z = 0$) and at $0 < t \leqslant t_i$ is replaced by the flux condition

$$-D(\zeta, t)\frac{\partial C}{\partial z} + V(0, t)C(0, t) - q_0/\theta(0, t) = 0 \qquad [109]$$

where $\zeta \to 0$ from the positive direction of z and the dimensionless concentration of the inflow solution is equal to unity. Actually, the value of $D(\xi, t) = D_m[\theta(0, t)] + \lambda[L(t)]V(0, t)$ is calculated as follows: for $L(t) < 13\lambda_{\max}$

$$\lambda = \lambda_{\text{pore}} + L(t)(\lambda_{\max} - \lambda_{\text{pore}})/(13\lambda_{\max}) \qquad [110]$$

while for $L(t) \geqslant 13_{\max}$,

$$\lambda = \lambda_{\max}$$

in which $\lambda_{\max} = 3$ cm (Biggar and Nielsen, 1976) and $\lambda_{\text{pore}} = 0.1$ cm (Bresler and Laufer, 1974). Furthermore, $D_m[\theta(0, t)]$ is obtained from (Bresler, 1973) $D_m(0, t) = D_0 a \exp[b\theta(0, t)]$, with $D_0 = 0.04$ cm^2/hr, $a = 0.002$, and $b = 10$.

The soil moisture retention curves that are adapted for the numerical computations are those of Equation [7], with the same values of β, ψ, θ_s, and θ_r, as for the approximate model. Hysteresis in the $\theta(\psi)$ relationship is not considered in the computations. The hydraulic conductivity function (Equation [6]) is also calculated as in the approximate model with the same deterministic η, K_s^*, m_Y, and σ_Y and the same values of K_s^i, $i = 1, 2, \ldots, N$. The computed results of $C_i(z, t; Y_i)$ $i = 1, 2, \ldots, N$ are substituted into Equations [108a], [108b], and [108c] to calculate the three central statistical moments of C. The numerical computations (see Bresler, 1973) were carried out with 90 depth increments. The size of each increment was constant with depth but changed according to the value of K_s^i. For example, it ranged from 0.05 to 2 cm for $q_0 = 0.5$ cm/hr, and from 0.1 to 4 cm for $q_0 = 6.5$ cm/hr. The minimum value of Δt was chosen to be 0.02 hr.

3. Model's Results: An Example

The results demonstrated in Figure 5 are again for the Panoche soil with m_Y, σ_Y (Equation [39]) and all other soil parameters as before (i.e., with $\lambda = \lambda_{\max} = 3$ cm, $K_s^* = 0.22$, $\theta_s = 0.43$, $\theta_Y = 0.05$, $\beta = 0.36$, $\eta = 2.59$, $\psi_w = 15$ cm) and three parameters characterizing the flow as in Figure 1 (i.e., $\theta_n = 0.1$ $R = q_0 = 0.5$ cm/hr, and $t_i = 24$ hours).

The capability of the approximate model to simulate deterministic field conditions during infiltration and redistribution is demonstrated in Figure 5 (A-1 and B-1, respectively), and it can be seen that the agreement is not so close. This suggests that the approximate piston-type solution for water flow and the corresponding approximate solution for $C(z, t)$ are not accurate if a particular deterministic value is taken for K_s and the other parameters. The picture is drastically improved when the statistical moments of C for the entire field are of interest. To illustrate this point, the distribution of the first central moment is given in Figure 5 (A-2 and B-2) and that of the second moment in Figure 5 (A-3 and B-3), in terms of coefficient of variation (CV) of C. It can be seen that the agreement between the two methods of computation

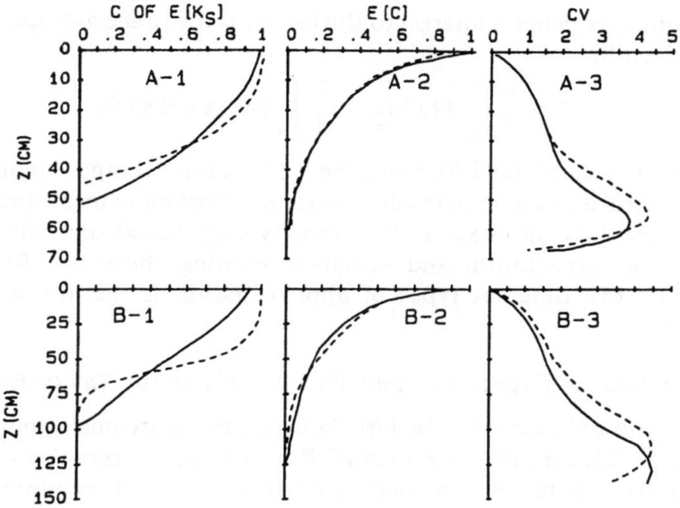

Figure 5. Computed profiles of concentration (C) as a function of depth (z) for $\lambda =$ 3 cm. A, B, 1, 2, 3, and water flow parameters are as in Figure 1.

is quite good for the expected values as well as for the coefficient of variation of C. The results given in Figure 5 also emphasize the dissimilarities between the shapes of $E[C]$ distribution with depth from numerical computations (and approximate computations) and the shapes of deterministic $C(z, t)$.

IV. Crop Yield

A. General Approach

In a field with a spatially variable soil parameter (such as K_s, η, β, θ_s, θ_r, etc.) and boundary and initial conditions (R or q_0, θ_n, C_n), crop yield Y (not to be confused with $Y = \ln \delta$ in Equation [12]) is regarded as a random function of the spatial coordinates of the field domain. The dependence of Y on the space coordinates (x, y) is indirect because crop yield depends generally on a set (vector) $\underline{R}$ of boundary and initial conditions (man-controlled variables) and on another set (or a vector) of spatial random functions $\underline{\alpha}$. Hence, Y is a functional depending on a set of random variables $\underline{\alpha} = \alpha^{(1)}, \alpha^{(2)}, \ldots, \alpha^{(n)}$, which are generally random functions of $\underline{x}(z, y)$ and the parameters $\alpha^{(i)}$ are characterized by their unconditional PDF. For any point in the field with area A (i.e., $x \in A$), crop yield depends on $\underline{R}(\underline{x})$ and $\underline{\alpha}(\underline{x})$,

$$Y(\underline{x}) = f[\underline{R}(\underline{x}), \underline{\alpha}(\underline{x})] = f[R^{(1)}(\underline{x}), \ldots, R^{(L)}(\underline{x}), \underline{\alpha}^{(1)}(\underline{x}), \ldots, \underline{\alpha}^{(n)}(x)] \quad [111]$$

The quantity of general interest to the farmer is the space average yield over the field given by

$$\bar{Y} = \frac{1}{A}\int_A Y(\underline{x})\,dx = \frac{1}{A}\int_A f(\underline{R}(\underline{x}), \underline{\alpha}(\underline{x})\,dx \qquad [112]$$

where A is the area of the field. Since the field spatial average $\bar{Y}$ is an integral, it can be regarded as a sum of random variables with finite mean and variance and therefore $\bar{Y}$ is approximately normally distributed and characterized entirely by its expectation and variance. Finding these two first central moments by two different types of approximation is the aim of the next subsection.

1. Computations of Expectation and Variance Using the Taylor Expansion

To derive approximate relationships between the statistical moments of $\bar{Y}$ (Equation [112]) and the moments of $\underline{R}(\underline{x})$ and $\underline{\alpha}(\underline{x})$, second-order Taylor expansion (see Abramowitz and Segun, 1964), about the expectations $E[\underline{R}(\underline{x})]$ and $E[\underline{\alpha}(\underline{x})]$, is used to give

$$\begin{aligned} Y(R^{(1)},\ldots,R^{(L)},\alpha^{(1)},\ldots,\alpha^{(n)}) &= Y\{E[R^{(1)}],\ldots,E[R^{(L)}],E[\alpha^{(1)}],\ldots,E[\alpha^{(n)}]\} \\ &+ \sum_{j=1}^{L}\left[\{R^{(j)} - E[R^{(j)}]\}\frac{\partial Y}{\partial R^{(j)}}\bigg|_{E[R^{(j)}]}\right] \\ &+ \sum_{j=1}^{n}\left[\{\alpha^{(j)} - E[\alpha^{(j)}]\}\frac{\partial Y}{\partial \alpha^{(j)}}\bigg|_{E[\alpha^{(j)}]}\right] \\ &+ \frac{1}{2}\sum_{j=1}^{L}\sum_{k=1}^{L}\Bigg[\{R^{(j)} - E[R^{(j)}]\}\{R^{(k)} \\ &- E[R^{(k)}]\}\frac{\partial^2 Y}{\partial R^{(j)}\partial R^{(k)}}\bigg|_{E[R^{(j)}],\,E[R^{(k)}]}\Bigg] \\ &+ \frac{1}{2}\sum_{j=1}^{n}\sum_{k=1}^{n}\Bigg[\{\alpha^{(j)} - E[\alpha^{(j)}]\}\{\alpha^{(k)} \\ &- E[\alpha^{(k)}]\}\frac{\partial^2 Y}{\partial \alpha^{(j)}\partial \alpha^{(k)}}\bigg|_{E[\alpha^{(j)}],\,E[\alpha^{(k)}]}\Bigg] \end{aligned} \qquad [113]$$

The expected value of $Y(\underline{x}) = Y[\underline{R}(\underline{x}), \underline{\alpha}(\underline{x})]$ is obtained by taking the expectation (averaging) of Equation [113] to get

$$\begin{aligned} E[Y] &= Y\{E[\alpha^{(1)}], E[\alpha^{(2)}] \ldots E[\alpha^{(k)}]\} \\ &+ \frac{1}{2}\sum_{j=1}^{N}\sum_{k=1}^{N} C^{(jk)}(\underline{x},\underline{x})\frac{\partial^2 Y}{\partial \alpha^{(j)}\partial \alpha^{(k)}}\bigg|_{E[\alpha^{(j)}],\,E[\alpha^{(k)}]} \end{aligned} \qquad [114]$$

where $N = L + n$ and α denotes hereafter any soil, man-controlled, and boundary and initial conditions, and $C^{(jk)}$ is the covariance defined by

$$C^{(jk)}(\underline{x}_1,\underline{x}_2) = E[\{\alpha^{(j)}(\underline{x}_1) - E[\alpha^{(j)}(\underline{x}_1)]\}\{\alpha^{(k)}(\underline{x}_2) - E[\alpha^{(k)}(\underline{x}_2)]\}] \qquad [115]$$

To simplify matters, it is assumed that $\alpha^{(j)}$ are weakly stationary and uncorrelated so that $E[\alpha^{(j)}] = \text{constant}$, $C^{(jk)} = 0$ for $j \neq k$, and $C^{(jk)} = \sigma^{(j)^2}$ = constant variance for $j = k$. Under these assumptions, the expectation of Y is

$$E[Y] = Y\{E[\alpha^{(1)}], E[\alpha^{(2)}], \ldots, E[\alpha^{(N)}]\} + \frac{1}{2}\sum_{j=1}^{N} \sigma^{(j)^2} \frac{\partial^2 Y}{\partial \alpha^{(j)^2}}\bigg|_{E[\alpha^{(j)}]} \qquad [116]$$

This means that the expectation of crop yield in a spatially variable field is approximately equal to the yield that would be achieved in a homogeneous field with the average parameter value plus the sum of products of the variance of each specific parameter and the second derivative of the yield, with respect to the specific parameter, evaluated at the expected value of that parameter. Equation [116] enables one to analyze the variability effect of each parameter on the expected yield.

Note that the second-order Taylor expansion may be a good approximation as long as $\sigma^{(j)^2}$ is relatively small for any soil or man-controlled parameter $i = 1, 2, \ldots, n$. This enables one to neglect even order terms larger than 2 in cases of multivariate normal distribution. In field cases where the variance of a given one or two parameters is relatively large and the other soil or control parameters are relatively uniform, crop yield expectation can be approximated by a numerical quadrature of

$$E[Y(\alpha^{(1)}, \alpha^{(2)})] = \int\int Y(\alpha^{(1)}, \alpha^{(2)}) f(\alpha^{(1)}) f(\alpha^{(2)})\, d\alpha^1\, d\alpha^2 \qquad [117]$$

if $\alpha^{(1)}$ and $\alpha^{(2)}$ are independent or by

$$E[Y(\alpha^{(1)})] = \int Y(\alpha^{(1)}) f(\alpha^{(1)})\, d\alpha^{(1)} \qquad [118]$$

if only one soil parameter is highly variable (the rest are deterministic) or if the influence of the variability of the other parameters on yield is negligibly small. Since $E[Y(\underline{x})]$ is independent of $\underline{x}$, then

$$E[\bar{Y}] = E[Y(\underline{x})] \qquad [119]$$

The variance $\sigma_{\bar{Y}}^2$ of field average yield $\bar{Y}$ (i.e., $\sigma_{\bar{Y}}^2 = \{\bar{Y}(\underline{x}) - E[\bar{Y}(\underline{x})]\}^2$), is given by

$$\sigma_{\bar{Y}}^2 = \frac{1}{A^2} \int_A \int_A C_Y(x_1, x_2)\, dx_1\, dx_2 \qquad [120]$$

where

$$\begin{aligned} C_Y &= E[\{Y(\underline{x}_1) - E(Y(x_1)]\}\{Y(x_2) - E[Y(x_2)\}] \\ &= E[Y'(x_1)Y'(x_2)] \end{aligned} \qquad [121]$$

is the covariance of Y. Taylor expansion of the residuals $Y'(\underline{\alpha}^{(j)})$ about $E[Y(\alpha^j)]$ yields the first-order expansion

$$Y'(\alpha^{(1)}, \alpha^{(2)}, \ldots, \alpha^{(2n)}) = \sum_{j=1}^{N} (\underline{\alpha}^{(j)} - E[\underline{\alpha}^{(j)}]) \frac{\partial Y}{\partial \alpha^{(j)}}\bigg|_{E[\alpha^{(j)}]} \qquad [122]$$

From Equations [119]–[122] one gets

$$\sigma_{\bar{Y}}^2 = \sum_{j=1}^{N} \sum_{k=1}^{N} \left\{ \left[\frac{1}{A^2} \int_A \int_A C^{(jk)}(\underline{x}_1, \underline{x}_2)\, d\underline{x}_1\, d\underline{x}_2 \right] \left(\frac{\partial Y}{\partial \alpha^{(j)}}\bigg|_{E[\alpha^{(j)}]} \right) \times \left(\frac{\partial Y}{\partial \alpha^{(k)}}\bigg|_{E[\alpha^{(k)}]} \right) \right\} \qquad [123]$$

For stationary $\alpha^{(j)}$

$$C^{(jk)}(\underline{x}_1, \underline{x}_2) = C^{jk}(\underline{x}_1 - \underline{x}_2) \qquad [124]$$

Using Cauchy algorithm (Feinerman et al., 1985)

$$\int\int C^{(jk)}(\underline{x}_1 - \underline{x}_2)\, dx_1\, dx_2 = \int C^{jk}(\underline{r}) H(\underline{r})\, dr \qquad [125]$$

where $\underline{r} = \underline{x}_1 - \underline{x}_2$, and $H(r)$ is the area of the domain covered jointly by A (the overlapping domain of A) and its translation by $\underline{r}$. Thus, if A is a square of sides D, $H(\underline{r})$ is given in a cartesian system of coordinates by

$$H(x, y) = \left(\frac{D}{2} - |x|\right)\left(\frac{D}{2} - |y|\right) \quad \text{for } |\underline{x}| < \frac{D}{2} \text{ and } |y| < \frac{D}{2}; \quad \text{otherwise, } H = 0. \qquad [126]$$

Similarly, for a circle of diameter D we have

$$H(r) = \frac{D^2}{2}\left[\cos^{-1}\left(\frac{r}{D}\right) - \frac{r}{D}\left(1 - \frac{r^2}{D^2}\right)^{1/2}\right] \qquad [127]$$

for $r = (x^2 + y^2)^{1/2} < D$; $H \equiv 0$ for $r > D$, while $H(r) = A$ for $r = 0$. Substitute Equation [125] into Equation [123] using Equation [124], and if $\alpha^{(j)}$ are uncorrelated functions then

$$\sigma_{\bar{Y}}^2 = \sum_{j=1}^{N} \left\{ \frac{\sigma^{(j)^2}}{A^2} \left[\int_A \rho^{(j)}(\underline{r}) H(\underline{r})\, dr \right] \left(\frac{\partial Y}{\partial \alpha^{(j)}}\bigg|_{E[\alpha^{j}]} \right)^2 \right\} \qquad [128]$$

where

$$\rho^{(j)}(\underline{r}) = C^{(j)}(\underline{r}) / \sigma^{(j)^2} \qquad [129]$$

is the autocorrelation function of the soil parameter or control variable $\alpha^{(j)}$. A simple autocorrelation function commonly used is the exponential

$$\rho^{(j)}(\underline{r}) = \exp[-r/I^{(j)}] \qquad [130]$$

where $I^{(j)}$ is the linear integral scale of the variable $\alpha^{(j)}$ defined by

$$\int_0^{\infty} \rho^{(j)}(\underline{r})\, d\underline{r} = I^{(j)} \qquad [131]$$

Hence, for an integral scale much smaller than the field "diameter" (D), that is, for $I^{(j)} \ll D$, $H(r) = A$ and Equation [128] becomes

$$\sigma_{\bar{Y}}^2 \cong \sum_{j=1}^{N} \frac{\sigma^{(j)2}}{A} \left[\int_A \rho^{(j)}(\underline{r})\, d\underline{r} \right] \left(\frac{\partial Y}{\partial \alpha^{(j)}} \bigg|_{E[\alpha^{(j)}]} \right)^2 \quad [132]$$

If, on the other hand, $\alpha^{(j)}$ are fully correlated, that is, $\rho^{(j)}(\underline{r}) = 1$ for any $\underline{r}$ and $\alpha^{(j)}$, then the contribution to $\sigma_{\bar{Y}}^2$ is

$$\sigma_{\bar{Y}}^2 = \sum_{j=1}^{N} \sigma^{(j)2} \left(\frac{\partial Y}{\partial \alpha^{(j)}} \bigg|_{E[\alpha^{(j)}]} \right)^2 \quad [133]$$

The modeling problem in order to calculate the two central moments of Equations [116] (or Equations [117] and [118]) and [132] or [133], boils down to the simulation of $Y\{E[\alpha^{(j)}]\}, j = 1, 2, \ldots, N$, and the estimation of the respective $\sigma^{(j)^2}$. First and second partial derivatives of Y with respect to each $\alpha^{(j)}$, evaluated at the expectation of the same parameter $E[\alpha^{(j)}]$, can then be obtained from the $Y(E[\alpha^{(j)}])$. This latter function can, in turn, be estimated from any transpiration model including salinity effects by assuming De Wit's relationships (Bresler et al., 1982) between transpiration and dry matter yield (Y)

$$Y = mT_r/E_0 \quad [134]$$

Here T_r is seasonal transpiration, E_0 is seasonal free water evaporation, and m is a crop coefficient. For a given crop and year with m and E_0 constant, relative yield is obtained from Equation [134] as

$$Y/Y_p = T_r/T_p \quad [135]$$

where Y_p and T_p are potential yield and potential transpiration, respectively. Deterministic models (Bresler et al., 1982) aimed at calculating transpiration (and therefore dry matter production or in some cases commercial yield) lead to the most general form of crop yield function such as (see Bresler et al., 1982):

$$Y = Y(\underline{\alpha}^{(j)}) = Y(\delta, K_s^*, \eta, \beta, \psi_w, \theta_s, \theta_r, \lambda, b, \psi_{cr}, \phi, m, F, E_0, R, C_0, \theta_n, C_n) \quad [136]$$

where δ, K_s^*, η, β, ψ_w, θ_s, θ_r, and λ are soil parameters characterizing water flow and solute transport; b, h_{cr}, ϕ, m, and F are plant parameters; R, C_0, θ_n, and C_n are man-controlled variables characterizing the boundary and initial conditions; and E_0 is a climatic boundary condition replacing R during the evapotranspiration stage. Here, b represents the area of influence of a root, ψ_{cr} is the critical (lowest possible) plant-root water potential, ϕ is the root distribution parameter, F is the crop factor, C_0 is solute concentration in water, and m, δ, K_s^*, η, β, ψ_w, θ_s, θ_r, λ, E_0, R, θ_n and C_n are as defined before.

The computation procedures for $E[Y(\underline{\alpha}^{(j)})]$ and $\sigma_{\bar{Y}}^2$ involve the following steps for each $\alpha^{(j)}$ (Equation [136]) for $j = 1, 2, \ldots, N$ and $N = 18$:

1. Input $\hat{E}[\alpha^{(j)}]$, $\hat{\sigma}^{j^2}, j = 1, 2, \ldots, N$ and divide the range between $\hat{E}[\alpha^{(j)}] + 2\hat{\sigma}^{(j)}$ and $\hat{E}[\alpha^{(j)}] - 2\hat{\sigma}^{(j)}$ (^ denotes estimated values) into $M - 1$ equal segments, such that $\alpha_1^{(j)} = \hat{E}[\hat{\sigma}^{(j)}] - 2\hat{\sigma}^{(j)}$ and $\alpha_M^{(j)} = \hat{E}[\alpha^{(j)}] + 2\hat{\sigma}^{(j)}$.

2. Set $i = i + 1$.
3. Calculate $Y_i^{(j)}\{E[\underline{\alpha}^{(k)}], \alpha_i^{(j)}\}$ $(k = 1, 2, \ldots, N; k \neq j)$ from numerical solution of Equation [136]. Note that for $N > j \geqslant 2$, $k = 1, 2, \ldots, j-1, j+1, \ldots, N$.
4. If $i < M$, go back to Step 2.
5. Apply spline subroutine and approximate

$$\left.\frac{\partial Y^j}{\partial \alpha^j}\right|_{E[\alpha^{(j)}]}, \qquad \left.\frac{\partial^2 Y^{(j)}}{\partial \alpha^{(j)2}}\right|_{E[\alpha^{(j)}]}$$

6. Calculate $\sigma_{\bar{Y}}^2$ from Equation [132] or [133] and $E[Y^{(j)}(\alpha^{(j)})]$ from Equation [116] (or [117] or [118], if applied, by a numerical quadrature over the range of $m_\alpha \pm 2\sigma_\alpha$).
7. Change j and repeat from Step 1 above.

B. Illustrative Examples

Examples of computed $Y^j\{E[\underline{\alpha}^k], \alpha^{(j)}\}$ for different univariant α are given in Figure 6, a–e. Also given in Figure 6 are the first and the second derivatives obtained by splines for the whole range of $\alpha^j = m_\alpha \pm 2\sigma$. Values of $E(\bar{Y})$ and $\sigma_{\bar{Y}}^2$ are readily obtained from Equations [116] and [133], respectively, when the derivatives are evaluated at the relevant $E[\alpha^{(j)}]$ values. Note that in the case of relatively large σ_α^2, and if univariant α is a good approximation, a numerical quadrature over the range given in each $Y(\alpha)$ line of Figure 6 can serve as a solution to Equation [118]. Similarly, $\sigma_{\bar{Y}}^2$ can be obtained by

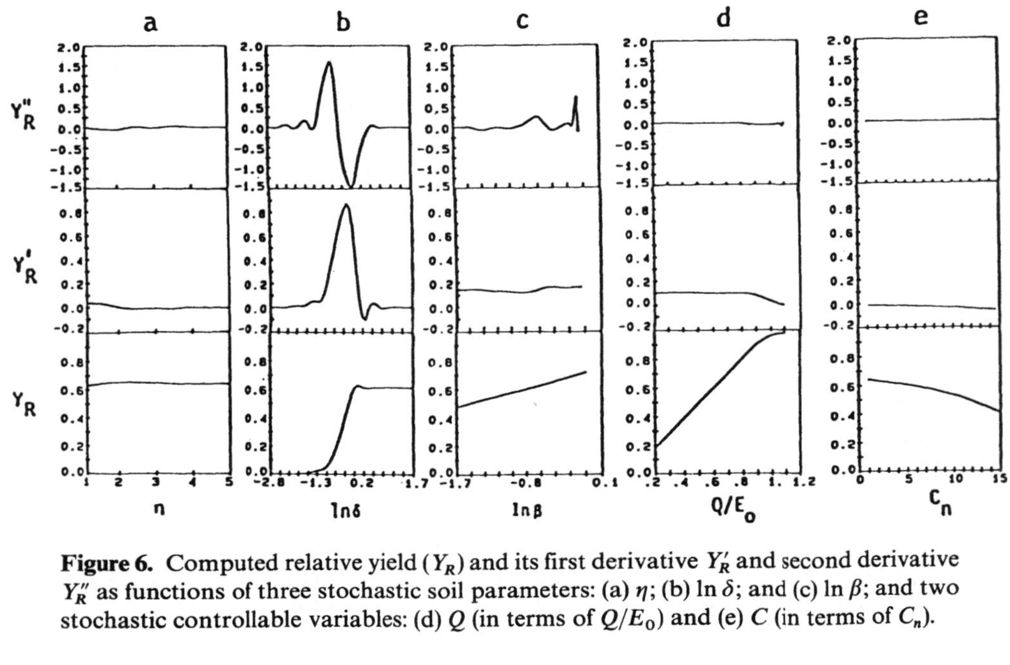

Figure 6. Computed relative yield (Y_R) and its first derivative Y_R' and second derivative Y_R'' as functions of three stochastic soil parameters: (a) η; (b) $\ln \delta$; and (c) $\ln \beta$; and two stochastic controllable variables: (d) Q (in terms of Q/E_0) and (e) C (in terms of C_n).

a numerical quadrature of

$$\sigma_{\bar{Y}}^2 = \int \{Y(\alpha) - E[Y(\alpha)]\}^2 f(\alpha)\, d\alpha$$

An additional possibility to derive an approximate relationship between the statistical moments of $\bar{Y}$ (Equation [112]) and those of $\underline{\alpha}^j(\underline{x})$, thus simplifying the solution for illustration purposes is to define a single man-controlled variable R and α as a scalar space function (not a vector of spatially variable functions). Alternatively, a complicated Monte Carlo simulation should be performed. Hence, for the simplified solution, let $Q = \int R(t)\, dt$ be a seasonal irrigation water quantity, the single man-controlled variable, and α be the scalar stationary-normal random function characterized by its constant mean $E[\alpha(\underline{x})] = m_\alpha$ and its two points covariance $C_\alpha(x_1, x_2)$ as given in Equations [115], [124], and [129] for $j = k = 1$.

To find the first central moments of $\bar{Y}$ ($E(\bar{Y})$, and $\sigma_{\bar{Y}}^2$) and with these simplifications for R and α, Equation [116] becomes

$$E(\bar{Y}) = f[Q, E(\alpha)] + \tfrac{1}{2} f^{(2)} \sigma_\alpha^2 \qquad [137]$$

and Equation [128] is expanded further such that a third-order Taylor expansion of $Y(\underline{x}) = f(Q, \alpha(\underline{x}))$ about $f(Q, E(\alpha))$ becomes

$$\sigma_{\bar{Y}}^2 = \sigma_\alpha^2 \left\{ (f^{(1)})^2 + \sigma_\alpha^2 \left[f^{(1)} f^{(3)} + \frac{1}{2} (f^{(2)})^2 + \frac{15}{36} (f^{(3)})^2 \sigma_\alpha^2 \right] \right\} \times \frac{1}{A^2} \int_A \rho_\alpha(\underline{r}) H(\underline{r})\, d\underline{r} \qquad [138]$$

Here, $f^{(n)}$ denotes the value of the nth derivatives of the production function $f = Y$ with respect to α, evaluated at $m_\alpha = E(\alpha)$.

Consider an agricultural field A hectares (ha) in size. The crop yield Y (t/ha) is assumed to be dependent on the deterministic quantity of water application Q(m^3/ha) and on a (two-dimensional) normally distributed random soil function $\alpha(\underline{x})$. The yield is related to Q and α by the Mitscherlich-type yield production function, (Mitscherlich, 1930).

$$Y(\underline{x}) = \alpha'[1 - \exp(-Q\alpha(\underline{x})] \qquad [139]$$

where the constant α' can be interpreted as the maximum possible yield and $\alpha(\underline{x})$ can be interpreted as a function depicting the yield spatial variation and may be related to the variability of various soil properties. In Equation [139], the spatial random function $\alpha(\underline{x})$ lumps all of the soil parameters influencing the efficiency of Q in its effects on Y.

To assess quantitatively the impact of uncertainty in $\alpha(x)$, without attempting to relate it to its origin, a specific case of the corn (Zea mays) crop is illustrated here with data of α' (maximum corn yield) of 25 t/ha taken from Stern and Bresler (1983).

For normal distribution of $\alpha(\underline{x})$, $E(\exp(-Q\alpha(\underline{x})) = \exp(-QE(\alpha) + \tfrac{1}{2}Q^2\sigma_\alpha^2)$

(see Equation [66] for $A \to \infty$), and hence the expected average yield over the field is given by:

$$E[Y(x)] = E(\bar{Y}) = \alpha'\{1 - \exp[-QE(\alpha) + \tfrac{1}{2}Q^2\sigma_\alpha^2]\} \qquad [140]$$

Equation [140] shows that $E(\bar{Y})$ increases with the average ($E(\alpha)$) and decreases with the variance (σ_α^2). The variance $\sigma_{\bar{Y}}^2$ is derived with the aid of Equations [120], [129], [139], and [140] and given by:

$$\sigma_{\bar{Y}}^2 = \alpha'^2 \exp(-2QE(\alpha) + Q^2\sigma_\alpha^2)\frac{1}{A^2}\int_A\int_A d\underline{x}_1\, d\underline{x}_2 \exp[\rho_\alpha(\underline{x}_1, \underline{x}_2)\sigma_\alpha^2 Q^2] - 1\} \qquad [141]$$

Hence, $\sigma_{\bar{Y}}^2$ increases with σ_α^2 and decreases with $E(\alpha)$.

For the autocorrelation function ρ_α of α, two extreme cases can be distinguished in Equation [141].

1. The values of α are not correlated, that is, $\rho_\alpha(\underline{x}_1, \underline{x}_2) = 0$, and in this case $\sigma_{\bar{Y}}^2 = 0$. In practice, such a situation may prevail when the integral scale (I) of α is very small compared with the field diameter (D). Intuitively, this result should be expected, as $\bar{Y}$ (Equation [112]) can be regarded as the sum of an infinite number of independent random variables, if the integral in Equation [112] is replaced by a sum.
2. The values of α are fully correlated, that is, $\rho_\alpha(\underline{x}_1, \underline{x}_2) = 1$. In practice, this situation may be assumed when $I \gg D$. In this case

$$\sigma_{\bar{Y}}^2 = \alpha'^2 \exp(-2QE(\alpha) + Q^2\sigma_\alpha^2)[\exp(\sigma_\alpha^2 Q^2) - 1] \qquad [142]$$

Again, on an intuitive basis, the existence of a correlation scale implies that $\bar{Y}$ is, roughly speaking, a sum of a finite number of random variables. In the extreme case $\rho_\alpha = 1$, the values of Y at all points are fully correlated and Y is a random variable independent of x.

For positive $0 < \rho_\alpha < 1$, Cases 1 and 2 give the bottom and upper limits of $\sigma_{\bar{Y}}^2$ so that

$$0 \leqslant \sigma_{\bar{Y}}^2 \leqslant \sigma_{\bar{Y}}^2 \qquad (\text{for } \rho_\alpha = 1)$$

Assuming, again for simplicity, a circular field and an exponential correlation function of the type $\rho_\alpha(r) = \exp(-r/I)$, and using the weighting function $H(r)$, Equation [141] becomes:

$$\sigma_{\bar{Y}}^2 = \alpha'^2 \exp(-2QE(\alpha) + Q_\alpha^2\sigma_\alpha^2)\frac{2\pi}{A^2}\int_0^D H(r)\{\exp[\exp(-r/I)\sigma_\alpha^2 Q^2] - 1\}\, r\, dr \qquad [143]$$

where, again, D is the diameter of the field, I is the linear integral scale of α defined by Equation [131], and $H(r)$ is the area of the overlapping domain of A and its translation defined by Equation [127].

Data on the variability of corn yield were used to estimate both $E(\alpha)$ and σ_α^2 using Equations [140] and [141], respectively, at $Q = 5{,}000$ m^3/ha and

Table 1. Field average expectation of irrigated corn yield, $E(\bar{Y})$, and its variance $\sigma_{\bar{Y}}^2$ calculated with "approximate" (Taylor expansion, Equations [137] and [138]) and "exact" (Mitscherlich, Equation [139]) production functions, for four seasonal water quantities (Q)

Q (m^3/ha)	$E(\bar{Y})$ (t/ha)		$\sigma_{\bar{Y}}^2$, $I' = 0.133$ $(t/ha)^2$		$\sigma_{\bar{Y}}^2$, $I' \to \infty$ ($\rho_\alpha = 1$) $(t/ha)^2$	
	Approx.	Exact	Approx.	Exact	Approx.	Exact
3,000	11.56	11.54	1.45	1.42	15.20	15.30
4,000	13.88	13.86	1.81	1.74	18.96	19.23
5,000	15.74	15.96	2.02	1.93	21.17	21.88
6,000	17.24	17.15	2.11	2.06	22.10	23.56

Variances were calculated for two values of integral scale (I) relative to field diameter (D) (i.e., $I' = I/D$).

Note: Data from Feinerman et al. (1985).

with $\rho_\alpha = 1$. The calculated values of the parameters $E(\alpha)$ and σ_α^2 are 0.0022 (ha/m^3) and 9×10^{-7} $(ha/m^3)^2$, respectively. The "approximate" (by Taylor's expansion) and "exact" (Mitscherlich) mean and variance of average yield $\bar{Y}$ for several levels of applied irrigation water (Q), for both $I' = I/D = 0.133$ and $I' \to \infty$ (i.e., $\rho_\alpha = 1$), are presented in Table 1.

The approximations of $E(\bar{Y})$ and $\sigma_{\bar{Y}}^2$ by a third-order Taylor expansion are quite accurate for the relevant range of applied water. This suggests that when the form of the production function does not enable one to derive exact expressions for $E(\bar{Y})$ and/or $\sigma_{\bar{Y}}^2$, approximations by Taylor expansion can be employed with confidence. It should be noted that when a second-order Taylor expansion was used, the deviations between the "exact" and "approximate" $\sigma_{\bar{Y}}^2$ were quite substantial for the relatively high levels of Q, and especially for $\rho_\alpha = 1$. As expected, the variance $\sigma_{\bar{Y}}^2$ increases with the ratio $I' = I/D$, reaching its highest value (for a given Q) for a fully correlated α (i.e., $\rho_\alpha = 1$ or $I \gg D$).

V. Economic Optimization

Any management process must eventually be judged from the point of view of economic efficiency. It is appropriate to include a discussion of recent modeling work on optimization. The aim of such a model is to compute the influence of spatial variability on economic optimization of one or more of the man-controlled variables, $R^{(j)}$, $(j = 1, \ldots, L)$ in Equation [111], or more specifically R or C_0 (or even C_n and θ_n) of Equation [136]. In the specific example given here, the impact of uncertainty upon the economic optimization of irrigation water quantity is emphasized. The problem is to estimate the effect of soil spatial variability, and of the resultant uncertainty in crop yield, on the policy of optimal and efficient use of water for irrigation, taking into account the farmer's attitude toward risk.

A. General Approach

The effect of risk on the farmer's behavior is expressed quantitatively by a utility function (U) defined in a univariant framework by a single random variable, the profit, (Π). The optimum level of the man-controlled variable R is the level that maximizes the expectation of the utility $E[U(\Pi)]$ rather than the profit in the deterministic case. The effects of uncertainty in $\underline{\alpha}^{(j)}$ on the optimal levels of $\underline{R}^j$ ($j = 1, \ldots, L$) are connected with two types of attitudes: (1) risk neutral and (2) risk aversion.

1. Risk Neutral

A farmer is risk neutral if he or she is indifferent to the certainty of the expectation of random profit and to the uncertain profit itself. This results in a linear utility function and therefore the expected utility is equivalent to the expected profit

$$E[U(\Pi)] = U[E(\Pi)] \qquad [144]$$

where a profit Π (in dollars per unit area of land) is selected as

$$\Pi = P_Y \bar{Y} - \underline{P}_R \underline{R} \qquad [145]$$

with P_Y being the price per unit of Y, and $\underline{P}_R$ the vector of the cost per unit of man-controlled variables $\underline{R}^{(j)}$. Using Equation [144] in Equation [145], the optimization problem is

$$\max_{\underline{R}} \{E(\Pi)\} = \max_{\underline{R}} \{E[P_Y \bar{Y} - \underline{P}_R \underline{R}]\} \qquad [146]$$

Conditions for maximum are: (1) first order

$$\frac{\partial E(\Pi)}{\partial \bar{R}} = P_Y \frac{\partial E(\bar{Y})}{\partial \underline{R}} - \underline{P}_R = 0 \qquad [147]$$

and (2) second order

$$\frac{\partial^2 E(\bar{Y})}{\partial R_i \partial R_j} < 0.$$

Note from Equation [112] that (1) $\bar{Y}$ is a random variable depending on the vector of random functions $\underline{\alpha}(x)$ and $R(x)$, and (2) in the risk neutral case, only the first moment $E(\bar{Y})$ has to be considered.

2. Risk Aversion

A farmer is risk averse if he or she prefers the certainty of $E(\Pi)$ to the uncertain Π (he or she will pay to avoid uncertainty) (i.e., concave utility function), so that

$$E[U(\Pi)] < U[E(\Pi)] \qquad [148]$$

The optimal solution for the risk-averse farmer is defined by

$$\text{Max}\, E\{U[AP_Y\bar{Y} - A\underline{P}_R\underline{R}]\} \qquad [149]$$

Again, first-order conditions for maximizing Equation [149] are

$$E\left\{\frac{\partial U(\Pi_t)}{\partial \Pi_t}\left[P_Y\frac{\partial \bar{Y}}{\partial \underline{R}} - \underline{P}_R\right]\right\} = 0$$

while second-order conditions are

$$\frac{\partial^2 E[U(\Pi_t)]}{\partial R_i \partial R_i} < 0 \qquad [150]$$

where $\Pi_t = A\Pi$ is the total profit from the field with area A.

B. Approximate Solution for Optimization

To derive approximate relationships between the moments of $\bar{Y}$ (Equation [112]) and $\alpha^{(J)}(\underline{x})$ and subsequently to solve the optimization problem, the simplifications that lead to Equations [137] and [138] are adopted.

The statistical moments of α are linked to the economic optimization models by Equations [137] and [138], which relate the mean and the variance of average yield to the level of the man-controlled variable, irrigation water (Q), and to the statistical parameters, $E(\alpha)$, σ_α^2, and ρ_α.

The optimization problem for the risk-neutral case is obtained by substitution of Equation [137] into Equation [146] to get

$$\underset{Q}{\text{Max}}\left\{P_Y\left[f(Q, E(\alpha)) + \frac{1}{2}f^{(2)}\sigma_\beta^2\right] - P_Q Q\right\} \qquad [151]$$

With first-order conditions

$$P_Y\left[\frac{\partial f(Q, E(\alpha))}{\partial Q} + \frac{1}{2}\sigma_\beta^2\frac{\partial^3 f(Q, E(\alpha))}{\partial E(\alpha)^2 \partial Q}\right] - P_Q = 0 \qquad [152]$$

and second-order conditions, decreasing expected marginal yield,

$$\frac{\partial^2 f(Q, E(\alpha))}{\partial Q^2} + \frac{1}{2}\sigma_\beta^2\frac{\partial^4 f(Q, E(\alpha))}{\partial E(\alpha)^2 \partial Q^2} < 0 \qquad [153]$$

To formulate the optimization problem for the risk-averse farmer, an exponetial utility function is adopted (Deaton and Muellbauer, 1980), that is,

$$U(\Pi_t) = -\exp(-\gamma A\Pi) \qquad [154]$$

where the constant (γ) is Arrow-Pratt measure of absolute risk aversion. Using Equation [154] in Equation [66], one gets

$$E\{-\exp(-\gamma A\Pi)\} = -\exp\{-\gamma AE(\Pi) + \tfrac{1}{2}\gamma^2 A^2\sigma_\Pi^2\} \quad [155]$$

and inserting Equation [155] into Equation [149] gives the optimization expression

$$\underset{Q}{\text{Max}}\left\{-\exp\left[-\gamma A(P_Y E(\bar{Y}) - P_Q Q) + \frac{1}{2}\gamma^2 P_Y^2 A^2 \sigma_{\bar{Y}}^2\right]\right\} \quad [156]$$

The first-order conditions for optima of the above expression are

$$P_Y \cdot \frac{\partial E(\bar{Y})}{\partial Q} - \frac{1}{2}\gamma P_Y^2 A \frac{\partial \sigma_{\bar{Y}}^2}{\partial Q} - P_Q = 0 \quad [157]$$

Second-order conditions for these maximization problems are given by the requirements of decreasing marginal yield or decreasing expected marginal yield, that is,

$$-\left[-\gamma A\frac{\partial E(\Pi)}{\partial Q} + \frac{1}{2}(\gamma A)^2\frac{\partial \sigma_\Pi^2}{\partial Q}\right]^2 - \left[-\gamma A\frac{\partial^2 E(\Pi)}{\partial Q^2} + \frac{1}{2}(\gamma A)^2\frac{\partial^2 \sigma_\Pi^2}{\partial Q^2}\right] < 0 \quad [158]$$

The optimization solution for the Mitscherlich-type production function (Equation [139]) and for the negative exponential utility function (Equation [154]) is, for the risk-neutral case:

$$\underset{Q}{\text{Max}}\left\{P_Y\alpha'\left[1 - \exp\left(-QE(\alpha) + \frac{1}{2}Q^2\sigma_\alpha^2\right)\right] - P_Q Q\right\} \quad [159]$$

with first-order conditions given by

$$P_Y\alpha'(E(\alpha) - Q\sigma_\alpha^2)\exp(-QE(\alpha) + \tfrac{1}{2}Q^2\sigma_\alpha^2) - P_Q = 0 \Rightarrow Q = Q_n \quad [160]$$

For the risk aversion case, optimization is given by

$$\underset{Q}{\text{Max}}\left\{-\exp\left\{-\gamma A\left[P_Y\alpha'\left[1 - \exp\left(-QE(\alpha) + \frac{1}{2}Q^2\sigma_\alpha^2\right)\right]\right.\right.\right.$$
$$\left.\left.\left. - P_Q Q\right] + \frac{1}{2}(A\gamma P_Y)^2\sigma_{\bar{Y}}^2\right\}\right\} \quad [161]$$

The first-order condition for optimum for this case is:

$$P_Y\alpha(E(\alpha) - Q\sigma_\alpha^2)\exp\left(-QE(\alpha) + \frac{1}{2}Q^2\sigma_\alpha^2\right)^2 - \frac{1}{2}\gamma P_Y^2 A\frac{\partial\sigma_{\bar{Y}}^2}{\partial Q} - P_Q$$
$$= 0 \Rightarrow Q = Q_a \quad [162]$$

The impact of changes in the three parameters (the ratio between the integral scale and the field diameter ($I' = I/D$), mean $E(\alpha)$ of α, and variance (σ_α^2) of α on the optimal levels of Q) is illustrated in the example (Table 2), using the basic parametric values of A, α', $E(\beta)$, σ_β^2, I', P_Y, and P_Q, which are: 1 ha, 25 t/ha, 0.0022 (ha/m^3), 9×10^{-7} (ha/m^3)^2, 0.133, 52.6$/t and 0.1$/m^3, respectively. The risk aversion parameter (γ) is varied between 0.001 and 0.01

Table 2. Optimal irrigation water quantity (Q) for irrigated corn, using "approximate" (Equation [137]) and "exact" (Equation [139]) production function for two parametric values of $I' = I/D$, one value of each $E(\alpha)$ and σ_α^2, and three values of γ.

	Q_c	Q_n(10 m^3/ha)		Q_a(10 m^3/ha)					
				$\gamma = 0.001$		$\gamma = 0.01$		$\gamma = 0.10$	
	(10 m^3/ha)	Approx.	Exact	Approx.	Exact	Approx.	Exact	Approx.	Exact
Base run	483	439	433	437	431	431	426	423	420
$I' = 0.266$	483	439	433	435	429	419	415	395	394
$I' \to \infty$ ($\rho_\alpha = 1$)	483	439	433	422	416	401	382	292	273
$E(\alpha) = 0.00242$	478	446	440	445	440	442	438	439	435
$\sigma_\alpha^2 = 1 \times 10^{-6}$	483	433	426	432	425	424	418	414	410

Water quantities for deterministic case, risk-averse farmer, and risk-neutral farmer are denoted by Q_c, Q_n, and Q_a, respectively.

Note: Data from Feinerman et al. (1985).

(Yassour et al., 1981). In addition, a sensitivity analysis is conducted by changing the values of $E(\alpha)$, σ_α^2, and I'. For $I' \to \infty$ ($\rho_\alpha = 1$), the assumption concerning the normal distribution of Π does not hold and the expectation $E[U(\Pi_t)]$ is calculated by a simple Monte-Carlo simulation.

Inspection of the "exact" results using the Mitscherlich equation (Equation [139]) shows that for the base run, optimal water application decreases by 11.5% for a risk-neutral grower. For risk-averse preferences, the range decreases by 12–15% for $I' = 0.133$, by 12.6–22.6% for $I' = 0.266$, and by 16.1–76.9% for $\rho_\beta = 1$. Hence, this example illustrates that optimal water applications are quite sensitive to the degree of risk aversity (γ) and to the ratio I/D (I'), and remain significant even when the grower is risk neutral ($Q_a < Q_n$ implies that water can be characterized as a marginally risk-increasing input). Water applications are relatively insensitive to changes in $E(\alpha)$ and σ_α^2, although there is a slight increase with $E(\alpha)$ and decrease with σ_α^2 of Q_n and Q_a.

The principal conclusion of the simulated data of Table 2 is that the impact of spatial variability and uncertainty of yield upon optimum level of water application might be substantial. In other words, economic prescriptions for optimal use of irrigation water that are based on the assumption that soils are perfectly homogeneous have the potential of being biased. This is the case whether relevant soil parameters are assumed to be known with certainty, or to be random, but their spatial variation in space is ignored (by assuming $\rho_\beta = 1$ for the entire field). This suggests that models focused on (1) formulation and estimation of crop-water production functions that include as arguments relevant soil parameters, and (2) estimation of the spatial variability of these parameters, have the potential to improve the ability of the farmer to select better irrigation policies.

The model presented here represents only the first step in a comprehensive assessment of the economic implications of heterogeneity of soils and of yield. Clearly, the analysis can be improved by examining the case where several relevant soil properties are simultaneously uncertain. Such an examination requires data on the multivariate joint distribution of these properties and estimation of a production function in which they are explicitly included as arguments. Beyond this, introduction of additional decision variables, such as the acreage devoted to irrigated agriculture (a variable that affects the variance of the average yield), where land has an opportunity cost, is an immediate extension. The impact of various agronomic technologies on the spatial variation characteristics may be a significant factor and is also deserving of further research.

List of Symbols

A	Area of a field
C	Dimensionless solute concentration
C^{jk}	Covariance between $\alpha^{(j)}$ and $\alpha^{(k)}$ variables

D_p	Effective molecular diffusivity
D_h	Mechanical dispersion coefficient
D	$D_h + D_p$
d_R	Deviation of R in rectangular PDF
$E(\alpha)$	Expectation of α
E_0	Free water evaporation
$f_\alpha(\alpha)$	Probability density function (PDF) of α
$H(r)$	Area of overlapping domain and its translation by r
I	Linear integral scale
K	Hydraulic conductivity
K_s	K at saturation
L	Wetting front
m	Crop coefficient
m_α	Mean of the variable α
$P_\alpha(\alpha)$	Cumulative probability of α
Q	Irrigation water quantity
Q_a	Optimum Q for the risk-averse farmer
Q_n	Optimum Q for the risk-neutral farmer
q	Vertical water flux (Darcy's flux)
q_0	Recharge, application rate (R)
$\underline{R}$	Vector of man-controlled random functions
R	Application rate (q_0)
r	Radial distance
S	Degree of saturation (reduced θ)
T_r	Transpiration
t	Time
t_i	Infiltration time
t_p	Ponding time
U	Utility function
V	Water velocity, or water depth
$\underline{x}$	Space coordinates' vector
x	Horizontal space coordinate
y	Horizontal space coordinate
Y	Crop yield (Sections III and IV)
Y	$\ln \delta$ (Sections I and II)
z	Vertical space coordinate
$\underline{\alpha}$	Vector of spatial random functions
$\alpha^{(j)}$	Specific spatial random function (soil property)
α'	Maximum possible yield
β	Constant coefficient in the $\psi(\theta)$ function
γ	Risk aversion constant
γ_α	Skewness of α
δ	Scaling parameter
λ	Dispersivity
μ	$\ln \lambda$

Π	Profit
η	Constant coefficient in the $K(\psi)$ function
ω	η/β
ρ	Correlation coefficient
σ_α^2	Variance of α
ψ	Suction head
ψ_w	Air entry value of ψ
θ	Volumetric water content
θ_n	Initial θ
θ_s	θ at saturation
θ_r	Residual θ

References

Abramowitz, M., and I.A. Segun. 1964. *Handbook of Mathematical Functions*. Dover Publishing, Inc., New York.

Biggar, J.W., and D.R. Nielsen. 1976. Spatial variability of leaching characteristics of a field soil. *Water Resour. Res.* 12:78–84.

Bresler, E. 1973. Simultaneous transport of solute and water under transient unsaturated flow conditions. *Water Resour. Res.* 9:975–986.

Bresler, E., H. Bielorai, and A. Laufer. 1979. Field test of solution flow models in a heterogeneous irrigated cropped soil. *Water Resour. Res.* 15:645–652.

Bresler, E., and G. Dagan. 1979. Solute transport in unsaturated heterogeneous soil at field scale: II. Application. *Soil Sci. Soc. Am. J.* 43:467–472.

Bresler, E., and G. Dagan. 1981. Convection and pore scale dispersive solute transport in unsaturated heterogeneous fields. *Water Resour. Res.* 17:1783–1693.

Bresler, E., and G. Dagan. 1983a. Unsaturated flow in spatially variable fields: 2. Application of water flow models to various fields. *Water Resour. Res.* 19:421–428.

Bresler, E., and G. Dagan. 1983b. Unsaturated flow in spatially variable fields: 3. Solute transport models and their application to two fields. *Water Resour. Res.* 19:429–435.

Bresler, E., and A. Laufer. 1974. Anion exclusion and coupling effects in non-steady transport through unsaturated soils: II. Laboratory and numerical experiments. *Soil Sci. Soc. Am. Proc.* 38:213–218.

Bresler, E., B.L. McNeal, and D.L. Carter. 1982. *Saline and Sodic Soils*. Springer-Verlag, New York.

Bresler, E., D. Russo, and R.D. Miller. 1978. Rapid estimate of unsaturated hydraulic conductivity function. *Soil Sci. Soc. Am. J.* 42:170–172.

Brooks, R.H. and A.T. Corey. 1964. *Hydraulic Properties of Porous Media*. Colorado State University, Hydrology Paper No. 3.

Carslaw, H.S., and H.C. Jaeger. 1959. *Conduction of Heat in Solids*. Oxford University Press, New York.

Dagan, G., and E. Bresler. 1979. Solute dispersion in unsaturated heterogeneous soil at field scale: 1. Theory. *Soil Sci. Soc. Am. J.* 43:461–467.

Dagan, G., and E. Bresler. 1983. Unsaturated flow in spatially variable fields: 1. Derivation of models of infiltration and redistribution. *Water Resour. Res.* 19:413–420.

Deaton, A., and J. Muellbauer. 1980. *Economics and Consumer Behavior*. Cambridge University Press, Cambridge, United Kingdom.

Feddes, R.A., E. Bresler, and S.P. Neuman. 1974. Field test of a modified numerical model for water uptake by root systems. *Water Resour. Res.* 10:1199–1206.
Feinerman, E., E. Bresler, and G. Dagan. 1985. Optimization of a spatially variable resource: An illustration for irrigated crops. *Water Resour. Res.* 21(6):793–800.
Haan, C.T. 1977. *Statistical Methods in Hydrology.* Iowa State University Press, Awes.
Hanks, R.J., A. Klute, and E. Bresler. 1969. A numeric method for estimating infiltration, redistribution, drainage and evaporation of water from soil. *Water Resour. Res.* 5:1064–1069.
Mein, R.G., and C.L. Larson. 1973. Modeling infiltration during a steady rain. *Water Resour. Res.* 9:384–394.
Miller, E.E., and R.D. Miller. 1956. Physical theory for capillary flow phenomena. *J. Appl. Physics* 27:324–332.
Mitscherlich, E.A. 1930. *Die Bestinun des dengenbedurfnissum des bodens.* Berlin Verlagsbuchhandlung Paul Parey.
Nielsen, D.R., J.W. Biggar, and K.T. Erh. 1973. Spatial variability of field measured soil-water properties. *Hilgardia* 42:215–260.
Reiniger, P., and G.A. Bolt. 1972. Theory of chromatography and its applications to cation exchange in soils. *Neth. J. Agric. Sci.* 20:301–313.
Russo, D., and E. Bresler. 1981a. Soil hydraulic properties as stochastic processes: I. An analysis of field spatial variability. *Soil. Sci. Soc. Am. J.* 45(4):682–687.
Russo, D., nd E. Bresler. 1981b. Effect of field variability in soil hydraulic properties on unsaturated water and salt flows. *Soil Sci. Soc. Am. J.* 45(4):675–681.
Russo, D., and E. Bresler. 1982a. A univariate versus a multivariate parameter distribution in a stochastic conceptual analysis of unsaturated flow. *Water Resour. Res.* 18:483–488.
Russo, D., and E. Bresler. 1982b. Soil hydraulic properties as stochastic processes: II. Errors of estimates in heterogeneous field. *Soil Sci. Soc. Am. J.* 46(1):20–26.
Saffman, P.G. 1959. A theory of dispersion in a porous medium. *J. Fluid Mech.* 6:321–349.
Saffman, P.G. 1960. Dispersion due to molecular diffusion and macroscopic mixing in flow through a network of capillaries. *J. Fluid Mech.* 7:194–208.
Segol, G. 1977. A three-dimensional Galerkin finite element model for the analysis of a contaminant transport in saturated unsaturated porous media. In: W.G. Grey, G.F. Pinder, and C.A. Brebbia (eds.), *Finite Element in Water Resources.* Penetech Press, London, pp. 2123–2144.
Stern, J., and E. Bresler. 1983. Nonuniform sprinkler irrigation and crop yield. *Irrig. Sci.* 4:17–29.
Warrick, A.W., G.J. Mullen, and D.R. Nielsen. 1977. Scaling field measured soil hydraulic properties using a similar media concept. *Water Resour. Res.* 13:355–362.
Yassour, J., D. Zilberman, and G.C. Rausser. 1981. Optimal choices among alternative technologies with stochastic yield. *Am. J. Agric. Econ.* 63:718–723.

Pedogenesis in the Grassland and Adjacent Forests of the Great Plains

D.W. Anderson*

I. Introduction

The most important characteristics of the soils of grasslands are their dark, organic matter-rich surface layers and their substantial contents of nutritive elements. These soils are an important resource in that much of the world's grain and livestock are produced on them. The extensive agricultural develop-

*Contribution No. R504 from the Saskatchewan Institute of Pedology, University of Saskatchewan, Saskatoon, S7N OWO, Canada.

Advances in Soil Science, Volume 7

ment is a consequence of not only the general absence of forest cover, making conversion to agriculture relatively simple, but the natural fertility and resiliency of the soils. The generally favorable characteristics are the result of the interactions among several components of the soil's environment or the ecosystem over time. These interactions have determined the nature and degree of soil development. Therefore, a knowledge of the basic processes of soil formation and the environmental factors influencing the processes is important. This body of research, the science of pedology, allows not only an understanding of the origin of the soils but permits predictions of the soil's response to changes in its environment. These considerations are particularly important at this time, in that many scientists show concern for a significant deterioration in the quality of the soil resource resulting from conventional tillage methods (Haas et al., 1957; Rennie, 1982). New strategies, such as conservation tillage, are being proposed to check the decline. Research and resolution of these problems requires a concept of soil that considers soil as functioning and dynamic, as well as a means of organizing the knowledge gained. It is important that soils are viewed as a system.

II. Grassland Soils as a System

A concept of soils as a system or systems has wide application both in a functional and in a dynamic sense. Functionally, soils are open, dynamic systems with many interdependent and complex reactions that are affected by many external processes and conditions. In an organizational sense, soils are part of a continuum in both time and space; it is necessary to define the object of study, the attributes or components that make up the object, and how the object of study relates to the remainder of the population at increasing levels of generalization.

Early concepts of soil, probably conditioned by the philosophies of biological science (Nikiforoff, 1959), viewed soil as an "independent" natural body with particular attributes that were a result of the action of various soil-forming factors at a geographical location. This concept of soil served soil science well in its early development and, although recognizing a progression from young to old soils, tended to regard soils as static. This was probably because the attributes of interest were those that changed very slowly with time. Later concepts, best illustrated by those of Jenny (1941), Nikiforoff (1959), Buol et al. (1973), and Smeck et al. (1983), consider soil as an open, dynamic system changing in response to rhythms in its present environment and conditioned by the environments and processes of its past. The physical, chemical, and biological processes in soils collectively represent what is called the pedogenic process (Nikiforoff, 1959). The products of these processes ultimately influence the nature of the soil itself and many of the properties of components of the soil's environment. The soil system, the plant system, and other systems such as the groundwater and the near-ground climate, are all constituent parts of ecosystems. However, it is necessary even within the soil system for a more refined integration of the objects or systems of interest.

An early idea, developed by Ellis (1931), recognized the soil profile as an individual object of study. Soil profiles could be organized in a hierarchical manner: initially grouped into associates (soil series) or groups of similar profiles of the same soil drainage class; then into associations that were made up of associates of similar geological origin or parent materials; followed by combinations that were of similar physiographic occurrence; and finally soil zones that grouped soils on a basis of a general similarity of climate and vegetation. This system did much to organize the expanding body of knowledge in soil science and proved useful in the mapping of soils at varying levels of detail.

A later concept that is similar, but somewhat more sophisticated, (Figure 1) is patterned on the level of integration concept in ecology (Rowe, 1961). Similar hierarchical arrangements have been discussed by others (Dijkerman, 1974; Smeck et al., 1983). This concept is particularly valuable in relating the increasingly detailed information obtained at the cell, particle, aggregate, or horizon level to the larger universes with which pedology must ultimately be concerned. Objects at different integrative levels are related in a functional

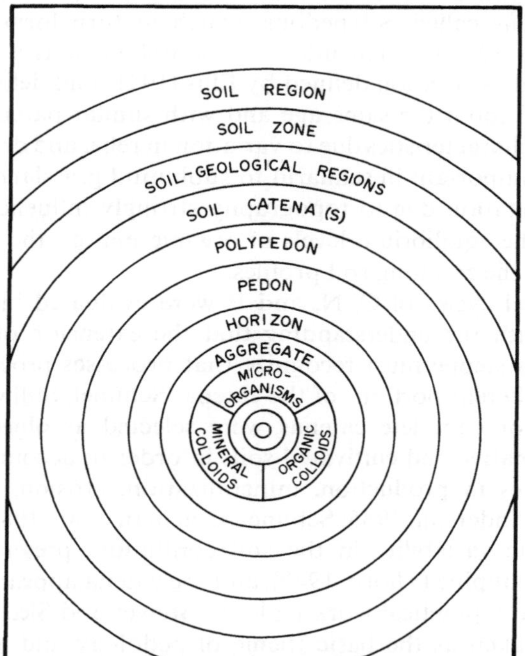

Figure 1. Objects of study in pedology at increasing levels of integration, or hierarchical levels.

sense, in that "the object of study of whatever level must contain, volumetrically and structurally, the objects of the lower levels, and must therefore be itself a part of the levels above" (Rowe, 1961, p. 421).

The usual object of study in pedology is the soil profile or pedon, which is composed of soil horizons. Soil horizons are the working aggregates of the whole system, and, like the organs of an organism, generally are adapted for the performance of some specific functions. Continuous performance of its specific functions imparts to each horizon its individual chemical and morphological character (Nikiforoff, 1959). Each horizon in turn is composed of natural arrangements of aggregates, peds, of particular morphology. Aggregates are an association of particles cemented by inorganic colloids and organic matter and are inhabited by microorganisms. A particular arrangement of peds, aggregates, and resulting pores does much to influence the supply of oxygen, water, nutrients, and energy to the microorganisms, and forms the environment through which plant roots grow.

A recent study of organic C, N, P, and S in soils of the Canadian prairies was based on the level-of-integration concept (Roberts et al., 1986). Detailed measurements were made on samples of horizons, but findings were related, systematically and appropriately, to ever-increasing universes by considering pedons, catenas, and soil zones.

At higher integrative levels, pedons are organized into groups of similar and contiguous pedons called polypedons, which in turn form parts of more general universes. Of particular interest in grassland soils is the soil catena, similar to the soil association defined by Ellis (1931) and defined as a related group of soils of about the same age and with similar parent materials but having different characteristics due to variation in relief and drainage. Catenas are particularly important in semiarid to subhumid grasslands. The redistribution of precipitation due to topography strongly influences productivity and, therefore, the equilibrium levels of organic matter, the water available for leaching and the resulting soil profiles.

Biogeochemical cycles of C, N, and P were evaluated in pedons within a soil catena, with the understanding that the extension of these findings to the shortgrass steppe must recognize that processes proceed to different degrees in the different portions of the catena (Schimel, Stillwell et al., 1985). Toposequences or complete catenas were selected as objects of study in comparisons of native and cultivated soils in order to account for the slope-related differences in production, mineralization, erosion, and deposition (Gregorich and Anderson, 1985; Schimel, Coleman et al., 1985).

The systematic variability in the soil·continuum presents problems of taxonomy and mapping (Thorp, 1947); and the catena appears to be a logical object of study at a practical working level. Brewer and Sleeman (1970) note that genesis is taken as the basic theme of pedology and has as its initial objectives the detailed study of soil properties, horizons, and pedons, and their logical arrangement into related groups by a classification system. However, studies are not complete without studies of soil distribution in the field so that the interrelationships among various kinds of soil in a landscape are known.

III. Classification of Grassland Soils

The logical arrangement of soils into related groups enables their properties to be remembered more easily and permits an organization of the knowledge about them. Therefore, classification is an important aspect of this discussion, particularly since the differentiae used relate to genetic properties or those important to soil development. This discussion is not intended to be comprehensive, but is simply necessary to later descriptions of genetic processes.

In early classification systems, the soils of the arid to humid grasslands were called Pedocals, soils in which calcium carbonate is an important constituent (Baldwin et al., 1938). Within this group, the zonal soils included Sierozem, Brown, Chestnut, and Chernozem, corresponding with a climatic-vegetation gradient that ranged from desert grassland through semiarid to humid grasslands (Thorp et al., 1948). This is similar to the present Canadian classification, except for minor differences in nomenclature and specific class limits in the Canadian system. The Food and Agriculture Organization (FAO) system corresponds generally with this sequence, with a grouping of the semiarid Brown and Chestnut soils as Kastanozems, the black soils of subhumid grasslands as Chernozems, with Greyzems in the forest transition (Table 1). These classifications all recognize an increasing degree of soil development from the semiarid grasslands to the forested regions. The groups formerly termed intrazonal soils, (those soils with special characteristics such as salt-affected or saline soils, solonetzic soils, and clayey (vertic) soils) are not included. These soil groups are discussed in greater detail later in this paper.

The present soil taxonomy (Soil Survey Staff, 1975) differentiates grassland soils into several classes based on soil temperatures and moisture conditions. These are properties that are related to other features of the profile, such as the color of surface horizons, the thickness and characteristics of other horizons, and the degree of soil development. Soils with dark surface horizons and a base-dominated exchange complex are considered to have a mollic epipedon and are grouped with the Mollisol Order. Mollisols are discussed in detail by Fenton (1983). Within the Great Plains, the cool soils of northern grasslands are termed Borolls (Figure 2), grading to Ustolls in warmer regions;

Table 1. A comparison of the FAO and Canadian classification and *Soil Taxonomy*, for soils of the northern Great Plains

Classification system	Great groups and suborders				
FAO	Haplic Kastanozem	Haplic Kastanozem	Haplic Chernozem	Greyzem	Albic Luvisol
Canadian	Brown	Dark Brown	Black	Dark Gray	Gray Luvisol
Soil Taxonomy	Aridic Boroll	Typic Boroll	Udic Boroll	Boralfic Boroll	Boralf

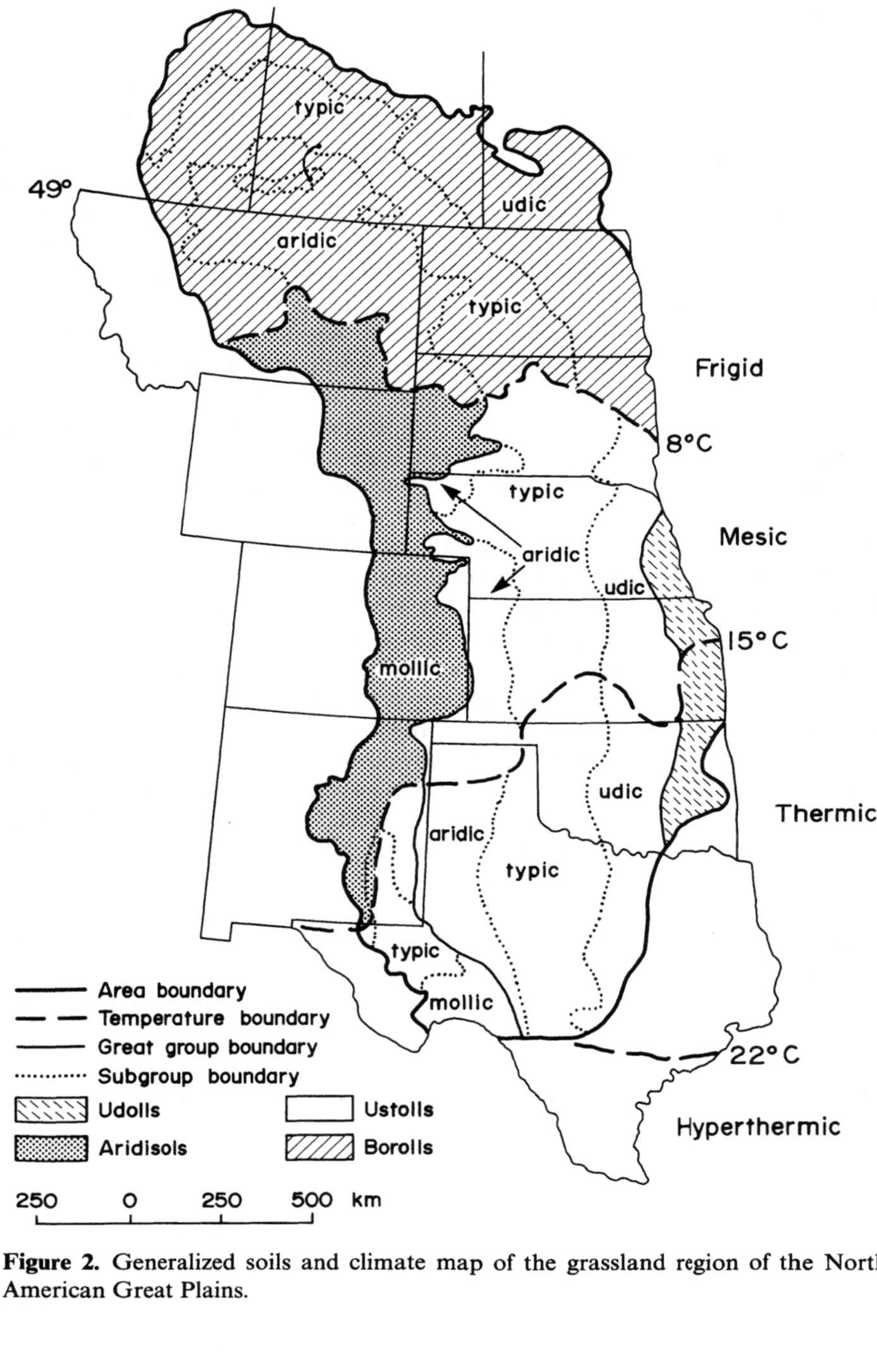

Figure 2. Generalized soils and climate map of the grassland region of the North American Great Plains.

Ustolls grade to Udolls with increasing moistness of the climate. The taxa conform in many ways to the great groups of the 1938 system but much more detailed differentiae are specified. The soils of grasslands with Mediterranean climates are Xerolls, occurring in areas with long, dry or xeric periods in summer. The lighter colored, organic matter-deficient soils of semiarid to arid grasslands are in the Aridisol Order. Solonetzic soils are differentiated as soils with natric or sodium-influenced B horizons within the Mollisols and Aridisols. The other major class dealt with in this review is the Vertisol Order. These are clayey soils with a dominance of expanding layer or smectite clays. These soils form wide and deep cracks upon drying, with considerable material falling down cracks, and major soil movement upon remoistening. This results in mixing of the soil to considerable depth.

IV. Soil Climates in Grasslands

In general, the climate of grasslands is one of extremes (Risser et al., 1981). The soil water content, a primary factor determining the nature and amount of vegetative cover, may be changing constantly because of precipitation patterns and winds that greatly influence the loss of water by evapotranspiration. Temperatures vary widely, both seasonally and diurnally. Grassland may occur in areas with extreme winter cold and where high summer temperatures and droughts are common.

Precipitation generally increases from west to east varying from as low as 250 mm at the shortgrass prairie-desert boundary to as much as 750–1,000 mm at the eastern transition from prairie to forest, and 500–650 mm in northern regions. As is evident on the generalized soils map (Figure 2), there is a strong relationship between soil formation and climate, particularly with temperature-precipitation interactions. Growing seasons are considerably shorter in northern regions, but precipitation is more effective because of cooler temperatures. The graphic depiction of the relation between great soil groups and climate (Arkley, 1967; Birkeland, 1984) requires the delineation of warm, mesic, and cool temperature classes to achieve clear-cut relationships. Another approach, using the biofactor that considers both temperature and precipitation (Kohnke et al., 1968), shows reasonable relationships as well (Figure 3). Several additional points were calculated for locations in the Canadian Plains and indicate that considerable differences in soil formation may result from relatively minor variations in moisture effectiveness where temperatures are cool. Some of the calculated points for locations with deep, organic matter-rich Chernozems (Borolls) fall in fields assigned to Gray Wooded (Boralf) soils in the original diagram. This may indicate that the biofactor, as calculated, requires modification for use in colder climates, or that the present climate is not consistent with the vegetation and soils now present. The latter is indicated by a general south and westward incursion of aspen forests into uncultivated grassland that occurred early in this century

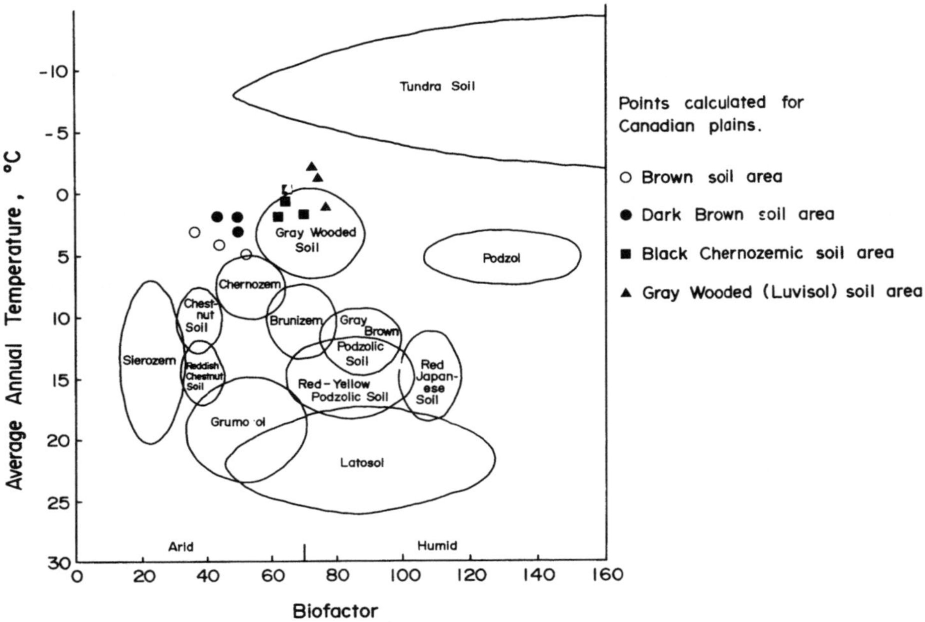

Figure 3. Climate-soil relationships using the biofactor (Kohnke et al., 1968). Additional points for the Canadian plains are shown. With permission of VCH Verlagsgesellschaft. Copyright © 1968.

when settlement resulted in a marked reduction in prairie fires (Bird, 1961). Fires may have maintained the forest-grassland transition further north than it normally would have been.

Although major climatic gradients have strongly influenced soils in a zonal way (Mitchell et al., 1944; and many others), local differences in microclimate related to topography may result in equivalent if not greater differences in soil formation over relatively short differences (Malo et al., 1974). Within undulating to rolling landscapes, upland sites are drier because of a loss of water due to runoff, and lower areas are more moist because of runon (Ellis, 1938). The actual amount of rainfall received on exposed hilltops may be slightly less, because many rainfall events are accompanied by strong winds (Anderson et al., 1975). Average wind velocities are considerably less in protected lower areas than on exposed uplands (Anderson et al., 1975), and snow tends to be blown from uplands to build up in adjacent depressions. North-facing and, to some degree, east-facing slopes have cooler soil temperatures and less loss of moisture by evapotranspiration because of reduced intake of solar energy (Southard and Dirmhirn, 1972; Anderson et al., 1975).

The replacement of grass with forest may result in a more moist microclimate in that winds at ground level are dampened, shading occurs, and snow is evenly distributed and melts more slowly with less runoff (Kittredge, 1948; Volobuev, 1964). These effects are thought to result in a greater leaching potential in forests than in grasslands with equivalent precipitation (Kohnke et al., 1968). However, a recent study of a forested site with Boroll soils (Sanborn and Pawluk, 1983) indicates that a rapid depletion of moisture by the forest vegetation, a high moisture-holding capacity of the forest floor, and interception by the forest canopy all act as buffers against infiltration of minor rainfalls and limit leaching potential.

V. Pedogenic Processes

The physical, chemical, and biological processes in soils collectively represent the pedogenic process (Nikiforoff, 1959), in that the reactions ultimately determine the characteristics of the resulting soil. Considerations of pedogenesis have concentrated generally on processes that determine morphologically significant properties, such as horizons of clay or sesquioxide enrichment, leached horizons, or A horizons of specific character. These considerations are important, but it is obvious that pedogenesis also influences functional properties of the soil, such as the ability to retain and recycle nutrients, the environment for the soil microflora, and the medium presented to a developing root. Nikiforoff considers the essence of genetic soil horizons to be their function rather than their morphology.

Soils are dynamic, open systems (Jenny, 1941; Smeck et al., 1983), as depicted in a conceptual way by Buol et al. (1973). Early ideas on soil formation considered omnibus processes that were not completely understood and often defined more in terms of the resulting soil profiles than in an operative manner. Simonson (1959) presented a generalized theory in which soils, after an initial step of parent material deposition, were considered to form by basic processes of addition, removal, transformation, and transfer. These ideas and the concept of soils as open and dynamic systems form the basis for more detailed discussions of pedogenic processes in grassland soils.

VI. Organic Matter Formation and Transformation

Dark, organic matter-rich surface horizons are characteristic of grassland soils. The nature of these horizons is important both to their differentiation from related soils of adjacent biomes and their performance in agriculture. Although it has been usual to consider primarily the addition or accumulation of organic matter, more careful scrutiny shows that processes of transformation, transfer, and loss are involved as well. The establishment and maintenance of a particular equilibrium level of organic matter depends upon three factors: the relative amount of biomass added, the rates of organic residue and organic matter decomposition, and the capacity of the soil to retain organic matter.

Many studies have shown that the organic matter content of soils increases with increased moisture supplies, presumably due to greater production of organic residues on the more moist soils (Jenny, 1941, 1980; Newton et al., 1945; Anderson et al., 1974). Aboveground production is strongly correlated with the amount of organic matter to 1 m depth (Lura et al., 1986). The factor most dominant in determining production in semiarid regions is available moisture.

Recent research encompassing a semiarid to humid moisture gradient in Montana illustrates the effect of moisture on production and organic matter (Figure 4). The dry areas have about 600 kg ha^{-1} of annual productivity and Aridisol soils that contain 80 t ha^{-1} of organic matter (Munn et al., 1978). Organic matter increases to 250 t ha^{-1} in moist areas that have annual productivity of 1,600 kg ha^{-1} and Boroll soils. A comparison of soil groups of various textures and topographic characteristics indicates that production increases by two- to threefold in going from regions with 250–350 mm of precipitation to those with 600–750 mm of precipitation in a west-to-east transect across the central plains with mesic soil temperatures (Aandahl, 1982). This corresponds with progressive increases in soil organic matter content, as documented in other studies (Jenny, 1941).

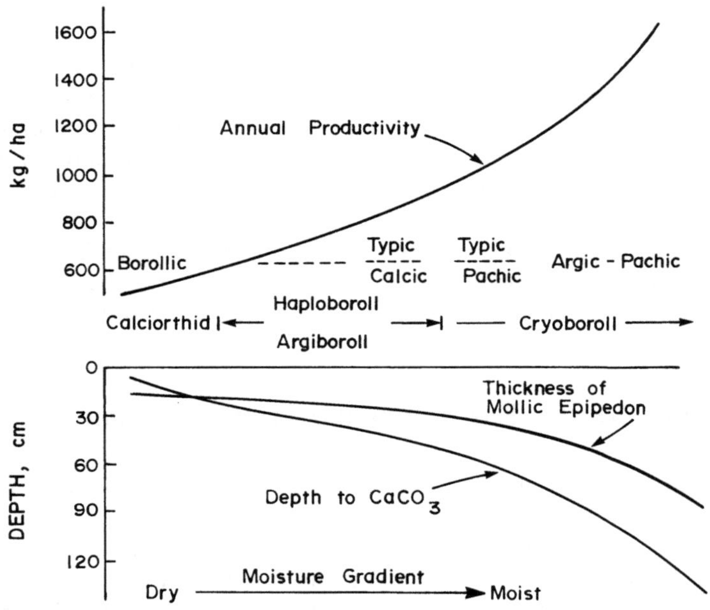

Figure 4. Soil-vegetation relationships along a dry to moist environmental gradient in the northern Great Plains (Munn et al., 1978). With permission of the American Society of Agronomy.

Although strong relationships between aboveground production and soil organic matter content are evident within similar temperature zones, below-ground production is a more important contributor to organic matter levels in grassland soils (Volobuev, 1964; Kononova, 1966). For example, on a loamy fine sand soil in North Dakota, aboveground production estimates are 2,400–3,020 kg ha^{-1} yr^{-1}, compared to below-ground production of 6,910–9,190 kg ha^{-1} yr^{-1} (Lauenroth and Whitman, 1977). Volobuev (1964) reports that the ratio between yearly decay of root substance to the yearly increment of above-ground production is about 5:1 for desert condition, 2:1 for dry grasslands, and less than 1 for forests. Below-ground production, although proportionately greater than aboveground in drier regions, increases in going from dry to moist regions, and corresponds with increases in soil organic matter content (Kolkhodzhayev, 1962).

Soil temperature and water content are interdependent in their effects on decomposition rates in soil (Volobuev, 1964). Variations in soil water tension influence decomposition rates more than variations in temperature (Nyhan, 1976). Maximum decomposition rates, however, combine moist soil conditions and moderate to high soil temperatures (Wildung et al., 1975). Analyses of data collected at the International Biological Programme (IBP) Grassland Biomes in the Great Plains show that decomposition rates increase with increases in incident solar radiation, presumably via its effect on soil temperature (Sims and Singh, 1978). The equilibrium level of organic matter is controlled by both input and decomposition rates, and generally decreases from cool to warm grassland environments (Nikiforoff, 1938; Jenny, 1980). This may be illustrated by a recent comparison of nine sites within the Great Plains that were selected to minimize variability in other soil environmental conditions (Figure 5). This 3×3 matrix shows that organic matter levels increase with increasing moisture within a temperature zone, and decrease with increased temperatures for comparable moisture conditions.

Moisture gradients similar to those occurring regionally within grasslands occur locally within hillslopes or catenas. Moisture conditions range from dry, runoff shedding upper slopes through moderately moist mid- or backslopes to locally moist lower slopes and concavities. This results in increases in biomass productivity and generally increased organic matter and N contents with distance downslope (Aandahl, 1948; Malo et al., 1974; King et al., 1983). Aboveground production of perennial grasses in a catena in the short-grass steppe increased from 702 ± 110 to 859 ± 110 to $1{,}160 \pm 338$ kg ha^{-1} in going from the ridge top to the footslope (Schimel, Stillwell et al., 1985). The production of annual or biennial forbs did not show a similar trend in the year of the study, which appears to have been an atypical year. Grain yields in the semiarid plains of Saskatchewan may be 40% greater on moist lower slopes than on knolls (Rennie and Clayton, 1967). Similarly, in a catena of Boroll soils in North Dakota, aboveground production of wheat increases from about 3 t ha^{-1} on knolls, to 4.5 t ha^{-1} in the gentle midslopes, and to

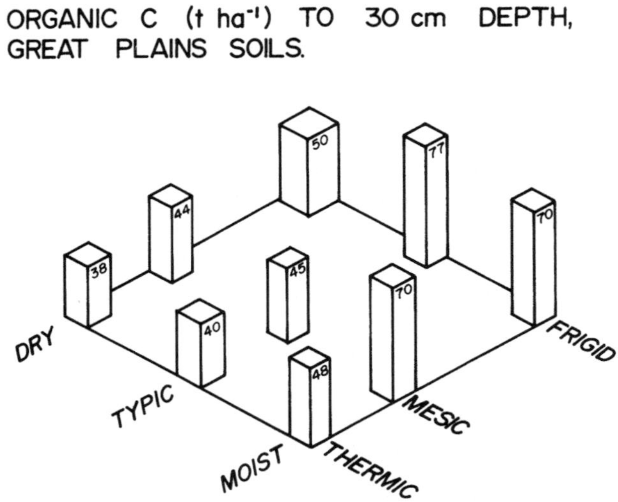

Figure 5. Organic carbon to 30 cm depth (t ha^{-1}) in Great Plain soils of different moisture and temperature regimes. (Unpublished data, Colorado State University.)

nearly 6 t ha^{-1} on the lower slopes where moisture supplies are greatest (Bauer et al., 1979).

Interactions between clay and organic colloids appear to be an important factor in the ability of many soils to retain organic matter and to recycle the nutrients temporarily tied up therein. The predominantly below-ground addition of residues as sloughed roots and root exudates permits humification to proceed in an environment in which mineral colloids are important interacting components.

Regression analyses of the large amounts of data gathered during soil surveys have enabled several researchers to examine relationships between organic C, clay content and precipitation. Organic C content was strongly correlated with the percentage of clay in 65 pedons of mesic and thermic temperature regimes in the Southern Great Plains (Nichols, 1984). McDaniel and Munn (1985), studying 143 grassland soils in Montana and Wyoming, reported a strong correlation between clay and organic C for mesic soils, and no relationship for frigid soils. Organic C was strongly correlated with precipitation and elevation, and not related to clay content, in 130 A horizons of cryic and frigid soils in Montana (Sims and Nielsen, 1986). Increases in plant production with increasing precipitation coupled with possible decreases in decomposition rates with increasing elevation explain the relationships that were observed. Clay appears to have a stronger influence on organic C contents in mesic and thermic soils where decomposition processes are stronger, and is less evident in soils where decomposition is dampened by cool temperatures. Interactions between clay and organic matter, however, may still influence the nature of organic matter and the cycling of nutrients.

Approximately one half of the organic matter in grassland soils is associated with the clay fraction, with a proportionately greater amount of associated nutrients such as *N* and *S* (Anderson et al., 1981). Some of this material is associated with the coarse clay and fine silt fractions and appears to be a strongly aromatic component of limited importance to nutrient cycling. Organic matter that is associated with fine clay, however, is much less aromatic and contains high concentrations of potentially labile humic compounds that are stabilized by sorption to clay. Radiocarbon dating of these materials indicates their importance to the medium-term cycling of energy and nutrients in grassland soils (Anderson and Paul, 1984). The complexing of organic matter and mineral components form the aggregates that are so important to the functioning of grassland soils, in that water and air entry to the soil is enhanced and root growth proceeds with little or no impediments. The intimate association of organic and mineral components (St. Arnaud and Whiteside, 1964; Dudas and Pawluk, 1969) in association with the dominance of calcium on the exchange complex may impede the translocation of clay from A to B horizons. Clay illuviation often leads to strongly structured subsoils that adversely affect moisture movement and root growth.

Changes in the soil environment influence both the quantity and qualitative characteristics of the soil organic matter. The organic matter of dry soils where biological activity is repressed due to drought is characteristically made up of partially decomposed residues and simple, relatively low molecular weight humic materials (fulvic acids), with minimal condensation of aromatic structures and considerable easily mineralized nutrient components (Volkovintser, 1969; Anderson et al., 1974). Increasing moisture supplies results in greater organic inputs, enhanced biological activity, and more organic matter. The proportion of weakly decomposed residues decreases and the relative amount of organic matter with large and complex molecules with strongly condensed aromatic structures (humic acids) increases. A greater proportion of the nutrients such as N and S are tied up in chemically resistant forms (Bettany et al., 1973, 1979; Anderson, 1979), particularly in soils of the more humid grasslands and adjacent forest. These observations indicate the importance of these pedogenic processes to nutrient cycling and soil fertility.

VII. Carbonate Weathering and pH

Processes involving organic matter dominate the development of grassland soils, but additional processes of weathering, transformation, and translocation are important as well. The significance of these processes and the factors controlling their rates can best be approached by considering soil properties along environmental gradients, both over long distances where regional climates are expressed, and along catenas where local factors condition the course and degree of soil development.

A study of grassland soils in Montana (Munn et al., 1978) showed that the depth of weathering of calcium carbonate ($CaCO_3$) increases from zero at the driest sites to 1 m or more at the most moist location (Figure 4). Detailed sampling transects along a soil catena in Saskatchewan indicates that the thickness of the solum or depth to $CaCO_3$ is least on locally dry knolls and generally increases downslope, corresponding with increasing moistness of the soil environment (King et al., 1983). Moisture-receiving concave sites are much more strongly leached than runoff-shedding convex areas. As the depth of $CaCO_3$ weathering increases, the pH of the surface horizons decreases, particularly in the thick soils in depressional areas.

Quantitative relationships between soil pH and climate, as expressed by the biofactor (Kohnke et al., 1968), indicate that soils are increasingly acidic with increasing moistness of the climate. Statistical relationships between base saturation, or the proportion of the exchange complex occupied by acidic cations such as aluminum or hydrogen, and climate indicate a general decrease in base saturation with increasing moisture. Improved statistical relations were obtained by stratifying soils on a basis of the acid or basic nature of the parent material, and grassland or forest soils. Base saturation values are considerably lower on acidic materials for grassland and forest soils. In relation to grassland soils, those of the forest have much lower base saturation values than would be predicted by the equation established in grasslands, indicating major differences in leaching and base recycling in the different biomes.

VIII. Clay Formation and Translocation

Transformations of phyllosilicate minerals as a result of pedogenesis are minimal in grassland soils (Allen and Fanning, 1983), presumably, because of the low intensity of weathering under relatively dry conditions. Soils in more humid portions of the grasslands and in adjacent forest may contain greater amounts of illite because of the fixation of K in a base-rich environment and slightly less smectite in surface horizons (St. Arnaud and Mortland, 1963). Most other studies of grassland and related soils indicate that most clay minerals are present in the same form as they occur in the parent material (Kodama, 1979; Sanborn and Pawluk, 1983).

A study of well and imperfectly drained Black soils in a glacial till landscape in the northern plains indicates that more intensive pedogenic weathering has taken place in the more moist soil (Huang and Lee, 1969). Under well-drained conditions, there is evidence of the transformation of vermiculite to mica. In the gleyed Eg horizons of the wetter soil, vermiculite is weathering to montmorillonite, kaolinite is being degraded, while quartz and amorphous aluminosilicates are accumulating. These observations indicate that the strongly acidic Eg horizon that is subject to variable redox conditions is a strong weathering environment. However, this kind of environment is more typical of forest than grassland soils.

Despite minimal weathering, many grassland soils show good evidence of the downward translocation of clay and the formation of argillic horizons.

Generally, the degree of clay translocation increases with increasing moistness and consequent leaching potential, both on a regional climatic basis and along catenas. A tabulation by Aandahl (1982) indicates that argillic horizons occur in soils on increasingly steep slopes as the climate becomes more moist, indicating the importance of moisture for leaching and weathering in their formation. The depth of solum and degree of development of argillic horizons increase from the summit to backslope and footslope in a catena in shortgrass steppe of Colorado (Figure 6) (Schimel, Stillwell et al., 1985). Soil properties such as organic C, N, and P also indicate increased weathering and biomass

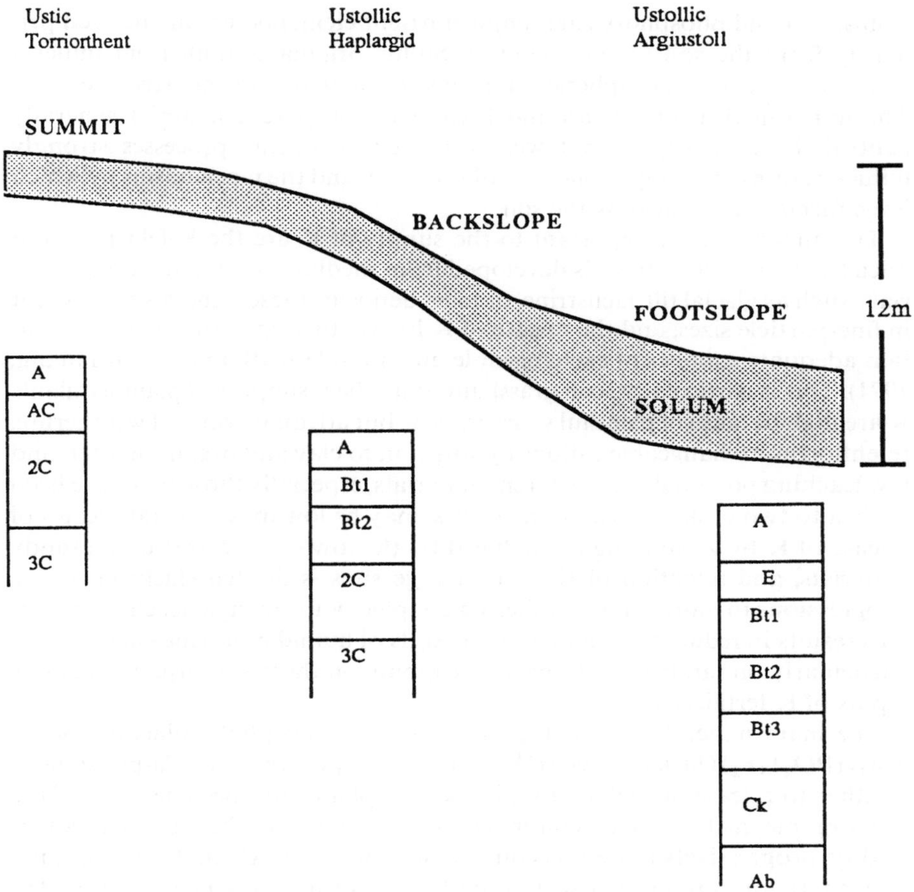

Figure 6. The relationship between topography and depth of solum in a grassland soil catena (Schimel, Stillwell et al., 1985). With permission of the Ecological Society of America.

production with increasing moistness downslope. Within catenas in areas of typic Boroll or Dark Brown soils, argillic horizons are general in the soils of locally moist lower slope and depressional areas (King et al., 1983). Solonetzic soils of grasslands have strongly developed B horizons with illuvial clay, and in situ clay formation is thought to be a factor in their development (Munn and Boehm, 1983). These properties result from soil formation in a sodium-affected environment, which is discussed in a later section.

IX. Pedogenesis and Major Nutrients

Potassium and phosphorus are major nutrient elements that are derived primarily from the soil parent material. Sulfur originates both from mineral sources and from atmospheric additions, particularly in industrial regions. The mineralogical properties and texture of the parent material generally control the total supply, but weathering or pedogenic processes strongly influence rates of supply in plant-available form and the mechanisms by which ionic forms are retained by the soil.

The minerals most important to the supply of K are the K-feldspars and mica (Jackson, 1964). In soils developed from medium- to fine-textured materials, such as glacial till, lacustrine, or loess deposits, these minerals are present in fine-particle sizes; and the release of K by weathering is sufficient to maintain adequate supplies of exchangeable and soluble K (Somasiri and Huang, 1971). This is particularly so in grassland soils where supplies of plant-available K are high because of plentiful sources, slow but adequate rates of weathering, retention as exchangeable cations by sorption to clay and organic matter, and low leaching potential. Sandy parent materials, especially those that have been subject to two or more weathering cycles, may be low in K minerals. Rates of release of K by weathering are reduced by the low surface area of the sandy materials, and retention of K on exchange sites is limited (Jackson, 1964). Progressions to more rapid weathering coupled with stronger leaching potentials results in reduced K supplies in the soils of humid grasslands and forests, particularly on sandy soils. Intensive cropping on these soils usually requires inputs of K fertilizers.

The main mineral form of P in soils is a Ca-phosphate mineral, apatite ($Ca_{10}(PO_4)_6F_2$) (Jackson, 1964). As pedogenesis proceeds, the Ca-phosphates weather to release phosphate that is used by plants and becomes part of the soil organic matter. With continued weathering and the development of acidity, progressively more P becomes associated with Al and Fe. In time, the phosphate ions are present within the lattices of hydrous Fe and Al oxides, or are coated by them, in the form termed occluded P.

Most grassland soils have about one half of their P associated with the relatively stable organic matter. The main sources of available P in these soils are from moderately labile inorganic forms, such as P adsorbed to amorphous minerals of Al and Fe, and the mineralization of organic P (Tiessen et al.,

1984). A comparison of the amounts and relative proportions of P fractions in the upper horizons of several Mollisol soils along an environmental gradient yields information on weathering processes (Westin and Buntley, 1967). The cooler and slightly drier Borolls have less Fe-associated P and more P associated with Ca than the Ustolls, indicating less intensive weathering in the Borolls. The Borolls have slightly more organic P and higher organic C:organic P ratios. There appears to be a higher rate of mineralization of organic P in the Ustolls.

Trends in the local distribution of organic and inorganic P from locally dry to moist portions of catenas were similar to those observed in comparing soils of semiarid to moist soil zones (Roberts et al., 1986). Concentrations of organic P increased from upper (locally dry) to lower slope positions, and from Brown to Black soils. The distribution of organic P was related to the transformations of inorganic P caused by differences in weathering intensity. The amount of Ca-phosphates in surface horizons decreased with increasing moisture.

Organic forms of S account for most of the S present in freely drained soils, with inorganic forms of S more important in weakly leached soils of more arid regions (Bettany et al., 1983; Maynard et al., 1984). Organic forms are most significant in soils where S is present in growth-limiting amounts, and interactions among C, N, P, and S are important (Stewart et al., 1983). Organic S in soils is generally divided into two fractions: the S bonded to C via ester linkages (C—O—S) and considered more biologically active, and S bonded directly to C in complex humic structures and considered much less active. A comparison of soils along a dry to moist environmental gradient in Saskatchewan showed that ester-bonded S is present in comparatively larger amounts in the drier soils with less intense humification processes, with progressively more C-bonded S with increased moistening of the soil environment (Bettany et al., 1973, 1979). Similar trends were observed for dry to moist gradients within catenas (Roberts and Bettany, 1985). Paralleling the increase in C-bonded S, there is an increase in the C:S ratio (a lower concentration of S in the organic matter), resulting in a much greater incidence of S deficiency in the soils of humid grasslands and forest regions. Assuming relatively similar amounts of S in the original parent material and similar atmospheric additions in preindustrial environments, it is evident that leaching losses of sulfate must be a factor in the more moist soils. Subsoil horizons containing gypsum ($CaSO_4 \cdot 2H_2O$) are common even in well-drained soils of semiarid regions, but rare in the soils with a greater leaching potential. This suggests deep leaching of sulfate S completely out of the soil profile in the more moist soils.

Grassland and forest soils may contain considerable amounts of organic S in their subsoils. A recent study has shown that this S is translocated to the subsoil by leaching when in soluble organic form (Schoenau and Bettany, 1987).

Variation in nutrient concentrations related to pedogenic factors may be discerned by consideration of the huge numbers of samples submitted to soils laboratories for analyses and fertilizer recommendations. In South Dakota, increasing precipitation results in reductions in P and K concentrations and soil pH, whereas soil organic matter content increases (Westin, 1976). Increased temperature lowers K levels, increases P, and has no effect on pH. Within a climatic zone, sandy soils are lower in K, P, and organic matter, with generally higher pH than finer textured soils. When $CaCO_3$ content increases, levels of P and K decrease and pH increases, with no effect on organic matter. Potassium, phosphorus, and organic matter are less on steeper slopes, and pH is higher, indicating loss of nutrients and weathered soil by erosion. Older soils have lower K contents and pH, with P and organic matter not affected by soil age. These findings illustrate the considerable impact of pedogenic factors and processes on soil performance in agriculture.

X. Processes in Clayey Grassland Soils

Grassland soils with a high content of expanding clays have distinctive properties that differentiate them from normal grassland soils, both taxonomically and functionally. These soils were originally called Grumusols, are Vertisols by the *Soil Taxonomy* (Soil Survey Staff, 1975), and are designated as special grumic families by the Canadian system (Canada Soil Survey Committee [CSSC], 1978). Sizeable areas of these soils occur in the Great Plains grassland. There are about 1.5 million ha in the Coastal Plains of Texas, where they occur on clayey alluvial and marine deposits of Pleistocene age (Kunze et al., 1963; Aandahl, 1982). They occur locally throughout the Great Plains on clayey deposits. They are common in the Canadian plains on clayey glacial lacustrine sediments, where they are considered among the best agricultural soils (Mermut and St. Arnaud, 1983).

The genesis of grumic or Vertisol soils is basically dependent on the expanding, churning, and cracking of a soil mass that is high in expanding clays (montmorillonite or smectite clays) and subject to pronounced changes in volume with change in moisture content. When dry, cracks develop that are a few to several centimeters across and up to 1 m or more deep. Surface material is displaced into cracks, mixing soil to considerable depth and increasing the subsoil volume. This leads to considerable stress when the soil re-wets and expands, resulting in the mass movement of one coherent mass of soil against another. Pressure faces, called slickensides, form where adjacent blocks slide against each other. Cracks tend to form around polygons, and, generally, soils crack in the same place each year (Lewis, 1977). There is a net downward movement of soil at the cracks coupled with upward displacement under the polygons caused by subsoil pressures on wetting. This results in an uneven surface configuration consisting of microknolls or microhummocks interspersed with microhollows, termed gilgai microrelief (Kunze et al., 1963; Soil Survey Staff, 1975; Graham and Southard, 1983). The soil movement involved with moisture change in Vertisols is sufficient to tilt trees and utility poles, and break foundations, highway surfaces, and pipelines (Anderson et al., 1973).

The mixing or churning of soil materials restricts the differentiation of soil horizons and the redistribution of soluble components that occur in less clayey soils. Minimal profile development is characteristic of the most clayey soils in semiarid regions, with a greater frequency of cracking, less leaching potential, and lower organic matter levels (Dasog et al., 1986). In slightly more humid regions, dark surface horizons with accumulations of organic matter are characteristic, as well as weak to moderate horizonation (Lewis, 1977; Mermut and St. Arnaud, 1983). The downward displacement of surface soil in cracks results in significant organic matter contents at depths of up to 1.5 m, dark bands, and layers (Mermut et al., 1983). Nutrients associated with the organic matter are distributed in a similar way, probably requiring different approaches to nutrient cycling, as compared to normal soils where nutrients are concentrated near the soil surface.

Micromorphological observations of clayey soils yield good evidence for stress phenomena and the churning process (Mermut and St. Arnaud, 1983). Granular aggregates typical of surface horizons are found mixed with angular fragments in the subsoil; planar voids and joint planes or miniature slickensides are also common.

Vertic subgroups in other classes in the taxonomy recognize soils that combine typical horizonation with characteristics such as relatively high clay contents, cracking, and slickensides (DeMent and Bartelli, 1969; Lewis, 1977). Gilgai are usually not present. Some Vertisols may form when argillic horizons become so enriched in clay through illuviation and in situ clay formation that they develop the shrink-swell potential of Vertisols.

Most soils wet evenly downwards, a result of the uniform infiltration of water over the entire soil surface. With Vertisols, rainfall that is received when the soils are dry and cracked may move downward through the cracks and result in an uneven distribution of moisture in the soil (Figure 7). Such a moisture distribution confounds normal ideas on moisture storage (Lewis, 1977), particularly those moisture simulation models that treat soil as a series of isotropic layers of different thicknesses parallel to the soil surface.

Precipitation received when these clayey soils are moist or wet infiltrates slowly, a consequence of the fine pores and the swelling of clays to seal the surface layer. This often results in surface ponding after heavy rains on level lands, or high rates of runoff on sloping areas. The surface ponding, coupled with a high moisture storage capacity that can sustain vegetation through droughty periods, results in plant communities that differ from those on loamy or sandy soils. For example, in the semiarid parts of the northern plains, distinctive wheat grass (*Agropyron dasystachyum*)–june grass (*Koelaria cristata*) communities are general on clayey soils, with wheat grass (*Agropyron* spp.) – spear grass (*Stipa* spp.) or wheat grass – blue grama (*Bouteloua gracilis*) common on soils of lower clay content (Coupland and Rowe, 1969).

On sloping clayey land, high rates of runoff resulting from slow infiltra-

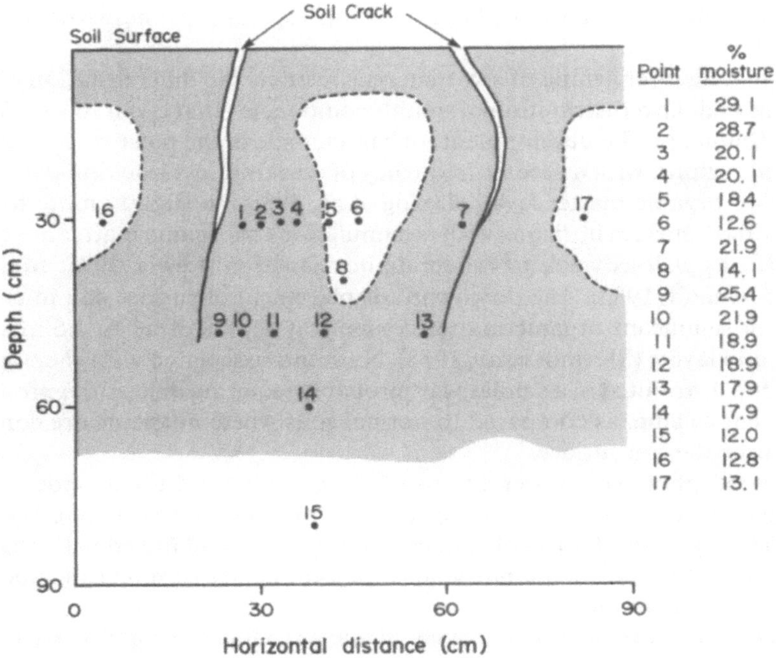

Figure 7. Moisture distribution in a vertic soil following 9 cm of rainfall. Shaded area indicates that part of the soil that has been moistened (Lewis, 1977). With permission of the American Society of Agronomy.

tion, however, may actually result in less moisture stored and lower herbage production than occurs on sandy, silty, or loamy soils (Aandahl, 1982). In addition, much of the moisture is stored in surface layers of clayey soils where it can be more easily lost by evaporation. Virtually all of the clayey soils are now cultivated in the northern plains. On sloping lands, erosion rates are substantial and have resulted in the displacement of large volumes of topsoil from upper to lower slopes (Mermut et al., 1983). This erosion has not substantially deteriorated soil quality in most cases, but is a potential problem where relatively thin clayey deposits overlay coarse substrates or saline subsoils.

XI. Solonetzic Soils

Extensive areas of the world's grasslands have solonetzic soils. These soils are important not only because of their areal extent and their peculiar properties so important to their use in agriculture, but because of the interesting interplay of local and far-distant environmental factors that are responsible for their development (White, 1964; Pawluk, 1982; Munn and Boehm, 1983).

Solonetzic soils have dense, clayey B horizons with high soluble and exchangeable sodium (Na) concentrations that are termed natric B horizons. The high Na levels result in dispersed soil colloids, making the horizons slowly permeable to water and air, and restricting root growth. These B horizons develop a columnar structure, which is slowly pervious when wet and extremely hard when dry.

Solonetzic soils originate as a result of a multiplicity of factors variable in both time and space. It is generally considered that the development of a solonetzic soil involves cation exchange processes in which Na dominates the exchange complex. Because Na salts are soluble and readily leached from soils even in semiarid climates (Anderson, 1977), processes to restrict deep leaching or move Na upward to the zone of soil formation are implicated in solonetzic soil development. Solonetzic soils often occur on lowlands adjacent to uplands within the Canadian plains, as documented by studies of hydrogeological conditions (MacLean and Pawluk, 1975); they are influenced by regional groundwater flows that move salts and Na into their zone of formation or retard the leaching of Na originally present. The upward flux of soluble constituents is considered to result from artesian or piezometric pressure where water tables are relatively deep, or capillary rise in the fine soil pores and tubules above water tables (Pawluk, 1982). The ultimate expression of the upward flow of soluble constituents is saline and sodic soils, where the forces are great enough to bring salts to the surface. Upward movement of salt-charged waters from deep glacial or bedrock aquifers is an important process by which soils in the Canadian prairies have become saline (Henry et al., 1985). Counterbalancing the upward flux is the downward leaching of soluble constituents due to gravitational water in saturated soils—water moving through the large pores and passages. This results in a tenuous equilibrium that maintains attenuated salt and Na concentrations at some level at or near the surface. These conditions may change with time, usually in response to climatic changes, changes in regional hydrology, or changes in land use. As demonstrated in the case of cultivation-induced saline seeps, summerfallow or the elimination of vegetation on upland areas reduces water use and adds to hydrological flows to adjacent lower land (Ferguson and Bateridge, 1982).

It is generally considered that solonetzic or sodic soils develop when Na ions move upward in greater concentrations than either Mg or Ca. The lower solubility of salts of Ca and Mg results in their precipitation in the lower parts of the profile when the soil dries, and Na dominates the exchange in the upper portion (Munn and Boehm, 1983). Under conditions of low-electrolyte (salt) content and with at least 10–15% of the exchange complex occupied by Na, dispersion of the organic and mineral colloids occurs (de Sigmond, 1938; Pawluk, 1982). The dispersed colloids are mobilized and, as demonstrated in a model system (Anderson et al., 1979), wash downward in gravitational water moving through large pores. This forms a clay-enriched, dark clay pan or natric horizon close to the surface. The formation of the slowly permeable B horizon, and anisotropic conditions for upward flow brought about by differences in clay content that are a result of clay displacement from the A to B horizons, limits upward movement of Na to the upper portion of the B horizon. The absence of capillary flow to replenish Na in the surface, coupled with hydrolysis reactions and the displacement of Na by Ca and Mg derived from weathering or biocycling, results in solonetz soils with a mildly acidic A horizon over a dense, clayey Na-dominated B horizon. Strong conditions for weathering are a consequence of temporary saturation above the B horizon resulting in the formation of bleached, acidic eluvial horizons that are characteristic of solodized-solonetz soils. The presence of significant amounts of pedogenically formed amorphous aluminosilicates and the mineral clinoptilolite (Ca-Al silicate) in the Ah and E horizons of a solodized-solonetz soil indicates the strong weathering conditions there (Spiers et al., 1984). The clinoptilolite was considered to have formed by the weathering of smectites.

The natric horizons will persist as long as Na moves upward by capillary flow to the top of the B horizon. However, if the hydrogeological conditions change so that Na no longer is replenished through upward flows, its concentration changes and the B horizons break down rather quickly. The breakdown of the natric horizon permits deeper leaching, makes the upper solum more acidic, and enhances edaphic conditions, resulting in greater biomass production. The result is a deep, strongly developed, and reasonably productive soil (Munn and Boehm, 1983).

The significance of the distance of a soil surface above the water table and the role of leaching to counterbalance upward fluxes of soluble constituents makes topography of considerable importance to the genetic development of solonetzic soils, particularly in semiarid climates where small amounts of runoff or runon can have considerable effects. Jenny (1941) discusses the interplay of topography and groundwater in soil formation. Solonetzic soil catenas in Saskatchewan were described by Bentley and Rost (1947). Typically, these catenas include both solonetzic and normal grassland soils (Ellis et al., 1967). The latter occur mostly on gently sloping upland areas that are furthest from the water table, with the upward movement of Na by capillary rise not reaching the surface horizons (Figure 8). The loss of moisture by runoff and the locally arid characteristics of upland sites limits leaching (St. Arnaud, 1979). This results in relatively thin soils, without B horizons, or with B horizons in which clay weathering and translocation are minimal, as illustrated by profile AMA (Table 2). The electrical conductivity (EC) values or salt concentrations and soluble Na percentages in the subsoil vary little with depth, indicating a quasi-static condition balancing minimum leaching and slight upward fluxes. The presence of a strong Ck or calcic horizon does suggest that both leaching and upward fluxes are active, but dampened in these profiles. Calcic horizons are general in this area.

More strongly developed soils such as the BKW pedon (Table 2) are general on nearly level uplands and on the middle to lower portions of very gentle slopes with less runoff and greater leaching potential than sloping sites. This Typic Argiboroll has an argillic B horizon, with evident clay skins and organic

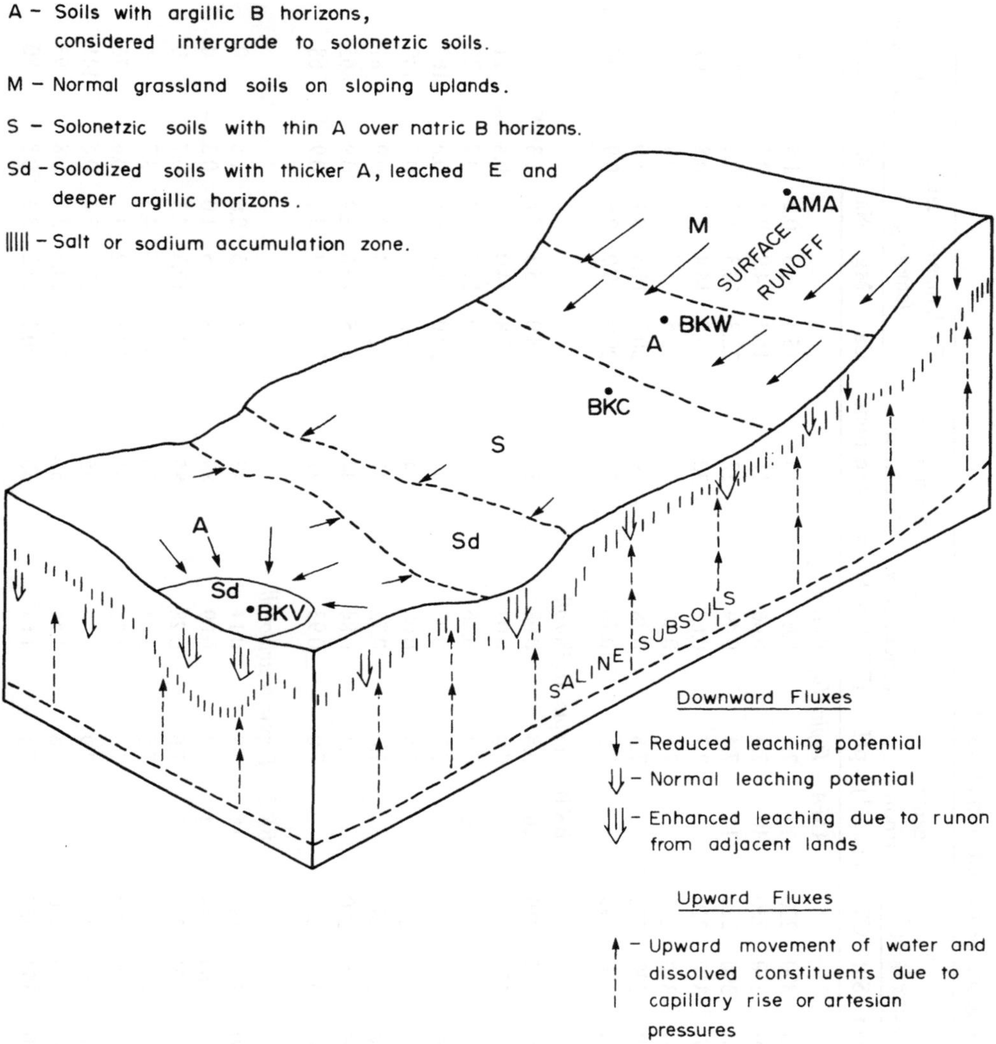

Figure 8. A catena of solonetzic and normal grassland soils in a semiarid region with regional upward hydrological forces.

staining on ped surfaces, and a distribution of clay that indicates clay movement from the A to B horizon. The maximum soluble Na concentration occurs in the upper C and then decreases with depth. This suggests that Na may have been once more actively involved in the pedogenic processes, particularly in promoting the mobilization of clay and organic matter. Exchangeable Na concentrations are low in the Bw horizon. These soils are designated as

Table 2. Properties of a catena including solonetzic and normal grassland soils

Horizon	Depth (cm)	Soil structure	Texture ($g\ kg^{-1}$) Sand	Silt	Clay	Organic carbon ($g\ kg^{-1}$)	pH	Electrical conductivity (EC)($S\ m^{-1}$)	Soluble Na % of total soluble cations	Exchangeable cations ($cmol\ kg^{-1}$) Ca	Mg	Na	K	$CaCO_3$ equivalent ($g\ kg^{-1}$)
						AMA – Typic Haploboroll								
Ap	0–13	Granular	450	330	220	14	7.4	0.07	10	15.3	3.3	0.2	1.5	0
BW	13–26	Prismatic	390	370	230	10	7.4	0.06	11	15.3	4.9	0.2	0.6	6
Ck	26–50	Massive	390	390	220	11	7.4	0.49	22	22.3	10.3	1.5	0.4	168
C	50–80	Massive	460	310	220		7.7	0.68	30	19.5	14.4	4.1	0.8	107
Cy1	80–105	Massive	450	310	250		7.6	0.86	30	22.5	11.5	3.3	0.9	67
Cy2	105–125	Massive	450	330	220		7.5	0.78	29	30.3	11.5	3.1	1.0	68
Cy3	125–150	Massive	500	310	190		7.5	0.76	30	20.0	9.9	2.6	0.9	66
						BKW – Typic Argiboroll								
AP	0–10	Granular	430	360	210	16	6.8	0.08	15	8.0	5.3	0.2	1.3	
Bt	10–26	Prismatic	390	300	310	7	7.0	0.07	32	12.8	14.0	0.6	0.8	11
BC	26–36	Prismatic	390	300	310	6	7.2	0.06	52	17.0	14.0	0.7	0.7	26
Ck	36–50	Massive	400	330	270		7.7	0.07	56	22.0	11.1	0.8	0.6	162
Cy1	50–70	Massive	400	340	260		7.8	0.08	74	21.3	12.7	1.4	0.7	140
Cy2	70–100	Massive	390	330	280		7.6	0.72	41	41.4	11.4	2.5	0.9	81
Cy3	100–125	Massive	380	340	280		7.4	0.84	44	83.5	10.6	2.9	0.9	65
Cy4	125–155	Massive	390	330	280		7.0	0.93	47	73.7	10.3	3.1	0.9	68
						BKC – Leptic Natriboroll								
Ap	0–13	Granular	450	370	180	15	5.8	0.11	47	6.3	4.5	0.6	1.5	0
E	13–20	Platy	560	360	74	4	6.4	0.09	89	1.8	2.5	1.0	0.3	0
Bnt	20–32	Columnar	350	400	253	7	7.6	0.28	85	4.3	12.3	4.3	0.7	0
Cy1	32–54	Massive	330	400	270		7.5	1.04	55	37.3	12.5	4.3	0.7	8
Cyk2	54–75	Massive	360	460	170		7.6	1.22	55	22.1	11.2	4.2	0.6	76
Cyk3	75–100	Massive	390	350	260		7.9	1.36	55	19.0	13.2	7.4	0.6	141
Cy4	100–125	Massive	400	330	270		7.8	1.54	51	25.1	14.5	3.5	0.8	93
Cy5	125–150	Massive	390	350	260		8.1	1.61	50	20.1	13.2	3.3	0.8	80

Table 2. Properties of a catena including solonetzic and normal grassland soils (*cont.*)

Horizon	Depth (cm)	Soil structure	Texture ($g\ kg^{-1}$) Sand	Silt	Clay	Organic carbon ($g\ kg^{-1}$)	pH	Electrical conductivity (EC)($S\ m^{-1}$)	Soluble Na % of total soluble cations	Exchangeable cations ($cmol\ kg^{-1}$) Ca	Mg	Na	K	$CaCO_3$ equivalent ($g\ kg^{-1}$)
						BKV – Albic Argiboroll								
Ap	0–10	Granular	460	340	200	18	6.7	0.12	7	10.3	4.9	0.3	2.1	0
E	10–18	Platy	530	320	150	7	7.0	0.06	8	6.8	3.3	0.2	1.4	0
EB	18–22	Prismatic	560	290	160	4	6.8	0.05	10	5.8	4.5	0.3	1.1	0
Bt	22–32	Blocky	390	280	330	4	6.8	0.05	22	8.8	15.2	0.4	1.6	6
Ck1	32–45	Massive	390	340	270		7.1	0.26	15	25.0	11.1	0.4	0.8	160
Ck2	45–60	Massive	400	340	270		7.2	0.59	28	32.0	11.4	0.9	0.9	170
Ck3	60–80	Massive	390	340	270		7.5	0.83	32	24.3	14.1	1.8	1.0	140
Cy1	80–105	Massive	390	320	290		7.5	0.86	36	111.1	12.6	2.2	1.1	80
Cy2	105–130	Massive	390	330	290		7.5	0.92	38	101.2	11.8	2.3	1.1	75
Cy3	130–155	Massive	400	330	270		7.5	0.90	38	31.7	11.8	2.7	1.2	70

Source: Unpublished data, Saskatchewan Institute of Pedology.

Solonetzic Dark Brown or intergrades to solonetzic soils by the Canadian soil classification (CSSC, 1978).

Soils with strongly developed, clayey B horizons, containing high concentrations of soluble and exchangeable Na, like the BKC pedon, are general through low-lying areas. It appears that the upward flux of Na is sufficient to maintain high Na concentrations in the B horizon. Soils like BKC occur mainly in lower areas where strong leaching due to runoff from adjacent slopes and strong upward fluxes interact in a dynamic way. Actually, the BKC pedon occurs in a toe-slope position and probably receives some extra water, resulting in the formation of the prominent E horizon and the slight degradation of the upper B that are typical of solodized-solonetz soils. Soils away from toe-slope retain the strong, natric B but lack E horizon development. Soils with strong natric horizons and high salt concentrations near the surface are termed Leptic Natriborolls.

Concave portions of the landscape that are imperfectly to well drained have deeply leached, strongly differentiated, profiles like BKV. Extra water resulting from runoff from adjacent lands moves the depth of maximum Na concentration well into the C horizon, particularly if the concavity, like the BKV site, occurs in an upland area. The BKV soil has a considerable concentration of exchangeable Mg in its B horizon, a characteristic of former solonetzic or natric soils that are being solodized (Soil Survey Staff, 1975). Many similar soils have dark-colored B horizons containing organic matter of illuvial origin relict from an earlier stage in development when Na was more active in soil formation (Anderson et al., 1979). The Canadian classification considers these soils as intergrades between solonetzic and chernozemic soils. They are Albic Argiborolls in that they have E horizons between the mollic surface horizon and the argillic or B horizon containing illuvial clay. Similar soils in downslope positions often have enough Na that the B horizons are designated as solonetzic or natric, resulting in their classification as Dark Brown Solods (Canadian) or Glossic Natriborolls.

The foregoing discussion describes soil catenas with systematic and generally predictable variability in soil properties over horizontal distances of a few to about 100 m, and vertical relief of 1/2–5 m. There are solonetzic soil landscapes with a similar variability at much smaller scale, including microdepressions or panspots with little or no vegetation interspersed with edaphically superior, well-vegetated soils on microknolls a few centimeters above the panspot (Munn and Boehm, 1983). Similar soil distributions have been discussed elsewhere (Mitchell et al., 1944). Factors of importance to the development of this fine-scale mosaic of soils include:

1. An underlying, sodic shale bedrock as a source of sodium and an influence on water flows.
2. Movement of water and soluble constituents, particularly sodium, from beneath the vegetated microknolls with good infiltration to the panspot where infiltration is much restricted by high exchangeable sodium levels. A net upward flux is maintained under the panspot.

3. Deflation by wind of A horizon material from the periphery of the panspot and its deposition on adjacent, well-vegetated microknolls. This further improves conditions for plant growth and infiltration on the microknoll.

These two discussions of solonetzic soil landscapes include working hypotheses that help to explain the variable distribution and highly dynamic nature of the soils. The importance of surface form, regional geohydrological conditions, near-surface stratigraphy, and vegetative cover are all important in determining the direction and status of soil development.

These considerations have practical importance in assessing the type and feasibility of reclamation practices on solonetzic soils. Certainly, it would be much more difficult to reclaim "active" solonetzic soils with strong upward flows of soluble constituents, as compared to "relict" solonetzic soils where the conditions contributing to their development no longer exist.

XII. Soils of the Grassland Forest Ecotone

A gradual increase in the organic matter content, enhanced leaching of bases, a decrease in the pH of surface horizons, and the potential for greater weathering and clay translocation coincides with increased moistness of the climate within grassland regions. As the grassland merges with forest, major changes in soil properties occur; sometimes gradually over relatively long distances in a wide transitional zone, and in other cases, abruptly over a few kilometers distance (Pettapiece, 1969). As noted earlier, the magnitude of the changes are greater than would be predicted by extending climate-soil formation relationships established in the grassland into forest regions (Kohnke et al., 1968). This indicates a major interaction between climate and forest vegetation. Forests may modify microclimates to make them more moist, although recent measurements question this, as discussed in the earlier section on climate (Sanborn and Pawluk, 1983). Other factors involve the nature and distribution of the vegetative residues added to the soil and the nature of the reactive organic components generated during its decomposition (Pettapiece, 1969).

Studies of soils in the transition zone in Illinois (Geis et al., 1970), further north in Minnesota (Severson and Arneman, 1973), and still further north in Saskatchewan (St. Arnaud and Whiteside, 1964) all have observed similar and rather abrupt changes in morphological and chemical properties. With the transition from grass to forest, there is a decrease in thickness and organic matter content of the surface mineral horizons and the buildup of a forest litter layer (Figure 9). This is a probable consequence of reduced below-ground inputs of organic residues and greater litter fall. Concomitant with reduced below-ground additions of organic residues, it is likely that decomposition rates of organic matter are enhanced by the generally more moist conditions. This results in gradual decreases in organic matter content, particularly in the lower parts of the A horizon. Reduced organic matter contents result in lighter colors in an incipient E horizon and less effective complexing of clay by humus

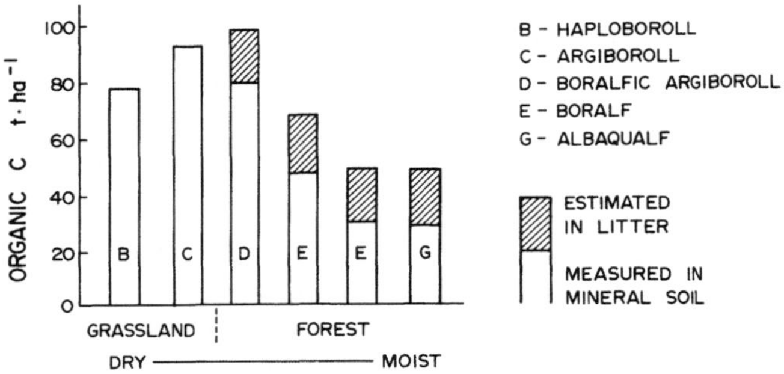

Figure 9. Organic carbon in soils along the grassland-forest transition in Saskatchewan. Estimates for litter horizons are based on a guideline established elsewhere that 1 cm of litter $\cong$ 10 t ha^{-1} of organic C. (Unpublished data, D.W. Anderson.)

(St. Arnaud and Whiteside, 1964), and may lead to increased translocation of clay downwards. Reactive organic acids generated by residues decomposing in the litter layer are thought to be involved in the greater leaching of clay and bases, and a shift to more acidic conditions (Pettapiece, 1969). With continued forest influence, A or Ah horizons become thin or absent, and strongly expressed eluvial or E horizons develop. A pronounced translocation of fine clay from the E to B horizon results (Figure 10), forming the argillic horizons that are typical of Alfisol or Luvisol soils.

A comparison of an otherwise similar Boroll soil of subhumid grassland with a Boralf soil of adjacent forest indicates that the degree of weathering is substantially greater in the Boralf (St. Arnaud and Sudom, 1981). Total losses of different minerals from the sola of the Boroll as compared to the Boralf are: Ca-feldspars, 20% versus 40%; Na-feldspars, 8% versus 22%; K-feldspars, 10% versus 17%. Calcium carbonates were leached below 23 cm in the Boroll, to 60 cm in the Boralf. These data indicate a significant degree of weathering in the grassland soil that is markedly greater just across the forest boundary.

Coincident with the changes in soil morphology and chemical properties, there are major differences in the amount, nature, and distribution of nutrients in grassland and forest soil systems. Those occurring mainly in organic forms, particularly N and S, are generally present in reduced concentrations in forest soils and appear to be tied up in resistant humic compounds (Bettany et al., 1973, 1979).

Proportionately large carbon and nutrient additions are made to the litter layer on the soil surface. Enhanced moisture conditions under forest, particularly in the litter layer, result in reasonably rapid decomposition rates and the production of simple, low molecular weight organic acids (or fulvic acids). These soluble organic components and their constituent nutrients, (particularly N, S, and P) may be translocated downward to the subsoil, essentially re-

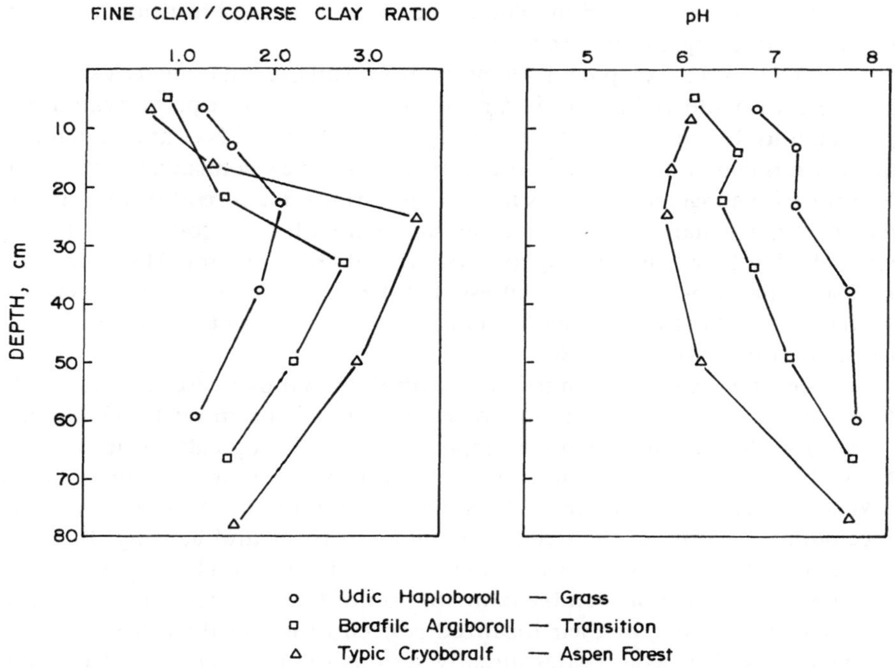

Figure 10. Fine clay: coarse clay ratios and pH for soils along the grassland-forest ecotone in Saskatchewan. (Unpublished data, D.W. Anderson.)

moving them from active cycling (Schoenau and Bettany, 1987). These essential differences in humification processes between organic litter layers and organic matter decomposing within mineral soil have been noted by others (McKeague et al., 1986). Where decomposition and subsequent humification occur mainly on the soil surface in the absence of fine mineral colloids, clay-humus complexes are not formed and have no role in nutrient processes. The medium-term protection of labile humic constituents by adsorption to clay is thought to be important to nutrient cycling in grassland soils (Anderson and Paul, 1984).

Nutrient elements that exist primarily as cations (such as Ca, Mg, and K) are generally present in lower concentrations in forest soils. This is a consequence of the enhanced weathering, greater leaching to depth, and the fewer cation exchange sites that relate to lower amounts of clay and organic matter in surface horizons.

Phosphorus, an element that varies considerably with stage of weathering (Smeck, 1973), changes both in nature and in absolute amounts. St. Arnaud (1986), using quantitative pedological methods, estimated that the weathering and removal of P from the E and Bt horizons of a Boralf soil under forest was substantial, whereas a Boroll soil of an adjacent grassland had lost little or no P during its formation. There had been a marked transformation of inorganic P to organic P in the Boroll.

Increasing soil development, albeit on moist soils, results in increases in Fe- and Al-phosphate, reductions in Ca-phosphates, and more plant available P (Smeck and Runge, 1971). The lower organic matter and N contents in forest and certain poorly drained soils does not appear to be a consequence of limited P supplies. Pedogenic processes may be important in concentrating elements like P in particular compounds at specific loci within the soils. A study of a strongly developed, imperfectly drained Gleyed Gray Luvisol (Albaqualf) soil indicates that two-thirds of the P present in the A horizon occurs in association with Fe, Al, and organic matter in localized concentrations termed concretions (Arshad and St. Arnaud, 1980).

A general observation of nutrient distribution in moving from grassland to forest ecosystems is that organic-associated nutrients move up (Hole and Nielsen, 1970) and cations move deeper into the soil. A greater proportion of the C, N, S, and other nutrients is included within the standing crop and litter layers in forest. Conversion of these lands to agriculture usually involves mechanical clearing of the trees, their piling in rows, and burning. In many cases, the littler layer is scraped off incidentally during the clearing operation. The net result is a considerable contribution of CO_2 to the atmosphere and the loss of nutrients, or their localized concentration in the ashes from the burned trees. This exacerbates already serious organic matter and nutrient deficiencies, particularly N and S, in cultivated Alfisol and Luvisol soils (Hoyt et al., 1978).

The foregoing discussion has dealt with comparisons in which major differences in soil properties are related to differences in vegetation at the grassland-forest boundary. This may not always be true. A detailed study of a Black (Boroll) pedon in the Canadian plains, that has probably been forested for 3,000 years, describes properties generally considered more typical of grassland than forest soils (Sanborn and Pawluk, 1983). The soil has a thick organic matter-rich A horizon, little evidence of clay translocation, and a high base status. Detailed process studies indicate relatively weak leaching and periodic wet or hydromorphic conditions. The lack of forest soil character may be a consequence of the lack of leaching, the rapid biocycling of bases, the imperfect drainage, or the stable clay-humus complexes in the A horizon that resist translocation of either component.

In contrast to simplified concepts of soil formation, it appears that a change in more than one factor, such as vegetation, may be required to bring about changes in soil properties. Hydrological conditions, on both a local or catena scale and at the regional level, appear to be an important interacting factor. In Illinois, after forest invasion, prairie soils with mollic surface horizons persist under imperfect drainage in lower sites; while more strongly differentiated Alfisol soils now occur on freely drained uplands (Geis et al., 1970). Pettapiece (1969) describes Dark Gray or Black soils under forest and considers them to be result of imperfect drainage.

A transect including 60 pedons under grass and aspen forest over gently rolling terrain showed that there are complex interactions among vegetation, topography, and soil properties (Anderson, 1986). Grassland occurred mostly on convex to planar upper slopes and had thin soil profiles, with shallow depths to $CaCO_3$ and somewhat saline subsoils. Forests dominated on concave and planar lower slopes and depressions. Forest soils had slightly more organic C in surface horizons, slightly lower pH, deeper profiles, and less saline subsoils than grassland soils. Forest soils on planar slopes had thick organic matter-rich A horizons and thin, often weakly calcareous Bw horizons. Well-developed E and Bt horizons occurred only under forest in concave, runoff-receiving lower areas. Topography, through its effect on redistributing moisture, appears to be as important as vegetation in determining the nature and intensity of soil development.

Comparisons in Saskatchewan of physiographic maps (Richards, 1969) with the generalized soils map (Moss and Clayton, 1969) indicate that strongly developed Luvisol (Alfisol) soils occur mainly on regional uplands that have a forest cover. In the southern portion of the boreal forest, Dark Gray or Black soils that are weakly leached and have thick, organic matter-rich A horizons occur under imperfect drainage conditions in lowland areas. These observations suggest that regional and local geohydrological conditions are important in determining the nature of soil formation, and may have greater influence than a related, and much more obvious factor, such as vegetation.

XIII. Effects of Man and Other Organisms

Organisms have considerable impact on processes in grassland soil systems. Fauna, particularly earthworms, play an active role in moving litter into the mineral soil or, alternatively, burying litter below surface-placed casts (Hole and Nielsen, 1970). Ants may bring subsoil from as deep as 2 m to the soil surface and cause modification of surface soil properties (Baxter and Hole, 1967). Burrowing rodents bring subsoil to the surface or carry topsoil down into tunnels in the subsoil. A Russian term, *krotovina*, is used to designate filled rodent tunnels in soil (Tyurin et al., 1959).

Soil fauna, primarily microarthropods, play a major role in the structural development and humus transformations in Boroll soils of the northern grassland and aspen parkland (Pawluk, 1985). Mixing of organic and mineral layers is minimal and likely reflects the general lack of fauna such as earthworms in the Boralf soils of the boreal forest region.

Large herbivores influence the development of grassland soils by grazing, thereby reducing the standing crop, slowing the buildup of a litter layer, and aiding the recycling of the nutrients excreted in feces and urine (Volobuev, 1964). Most herbivores do not graze evenly, and because most excretory products are deposited in resting areas, redistribute nutrients within grassland ecosystems (Woodmansee et al., 1981). Minimal losses of NH_3 by volatilization from feces or urine indicate that herbivores actually may conserve N, by converting grass to more easily decomposed forms and reducing losses from senescing vegetation and litter layers (Schimel et al., 1986).

Man and his technology have had a major influence on grassland soils, particularly in recent decades. Early man in the plains grasslands had little effect other than the burning of the prairie, which was done to stimulate the subsequent year's production and attract bison to the area (Warkentin, 1969). There were also many fires of natural origin. Hind (1859) explored the aspen parkland of the Canadian prairies and reported that fire was the main factor limiting the extension of forest into the grassland in this particular region. He considered that controlling the fires would permit the invasion of the prairie by aspen forest. The reduction in fires that accompanied the settlement of land did result in a considerable southward extension of the aspen forests into grassland (Bird, 1961). The effect of this forest invasion on soils varied from place to place. In some areas, particularly uplands that are well drained, a fairly rapid degradation of grassland soils to leached, Gray Luvisol soils appears to have occurred, substantially reducing soil quality for agriculture (Mitchell et al., 1944). In other areas, forest invasion appears to have had little effect on soil properties.

The conversion of grassland to cultivated agroecosystems has had substantial effects on soil properties. The application of inappropriate cultivation methods brought from Europe and eastern North America was a big factor in the catastrophic soil erosion of the 1930's dust bowls. These problems have largely been overcome by the introduction of soil-conserving tillage methods such as stubble mulch tillage (Fenster, 1977; Bauer and Black, 1981), but slower, and potentially as serious, processes still impact on the soils. The adoption of crop-fallow rotation systems has resulted in substantial declines in organic matter and nutrients and a deteriorating soil quality (Doughty et al., 1954; Haas et al., 1957; Rennie, 1982). The decline in soil organic matter has resulted from lower inputs of residues, particularly those added below ground, in cultivated systems and enhanced conditions of temperature and moisture that increases the rates of decomposition of organic matter in bare fallow fields. Reduced organic matter reduces infiltration of moisture, increasing runoff and erosion and further deteriorating the quality of the soil (Voroney et al., 1981; Tiessen et al., 1982).

In northern regions, mostly with glacial till soils overlying less permeable substrates, an increase in saline soil areas has been reported in recent years. The crop-fallow system of farming maintains the surface free of live vegetation every other growing season. In wet years when more precipitation falls than the soil can store, water moves below the root zone, increases the deep leaching of salts, and adds salts to the groundwater flows. Saline seeps occur when the groundwater flows more closely to the surface in lowland areas (Ferguson and Bateridge, 1982). Water seldom moves below the root zone under native vegetation.

Farming systems are being adapted to confront the problems of decreasing organic matter levels and increasing salinity. Generally these systems mimic natural systems by growing crops every year, reducing disturbance of the soil by cultivation, and making the most effective use of rain and snow. These systems are highly dependent on herbicides for weed control and require considerable research to demonstrate their effectiveness, their impact on the environment, and their economic potential.

XIV. Trends in Pedological Research

Research in soil genesis has built on the initial concepts of Dokuchaev (1983), who recognized that climate, subsoil (parent material), vegetation, fauna, man, age of the land surface, and relief are the significant pedogenic factors (Crocker, 1952). Pioneer work by Jenny (1941) made significant advances by proposing that the soil system could be described quantitatively by a series of equations relating any soil property to the soil-forming factors thus,

$$s = f(cl, o, r, p, t \ldots .)$$

where cl, o, r, p, and t represent climate, organisms, relief, parent material, and time, respectively. Despite the fact that this equation has not actually been solved in most cases, the approach resulted in methods of study in which the impact of one factor is assessed by keeping all other factors constant, allowing the factor of interest to vary in a regular manner and measuring the resulting effect on soil properties. This method has been gainfully used in many studies already mentioned in this chapter, and results in not only documentation of properties along environmental gradients, but can result in working hypotheses of the processes involved (Anderson, 1979).

The ecosystem, composed of soil, plant, and other involved systems, such as groundwater flows, becomes the object of study to those interested in the dynamics of soil systems (Crocker, 1952). Simulation modeling, a logical extension of the mathematical modeling of Jenny (1941), is a powerful tool in the study of these systems, to the pedologist and ecologist alike. Trudgill (1977) discusses simulation modeling of soil-plant systems and lists the following requirements for establishing linkages between components and the building in of more detail to general or conceptual models:

1. Research should study the effects of individual components on each other and their collective effect on the ecosystem.
2. Components most important in controlling the overall system must be identified and distinguished from less important ones.
3. The external factors that influence the state of the system should be identified and their effects quantified as much as possible.
4. The researcher should be able to predict the future state of the system in relation to external events and, ultimately, identify how man's actions may have an impact on the system.

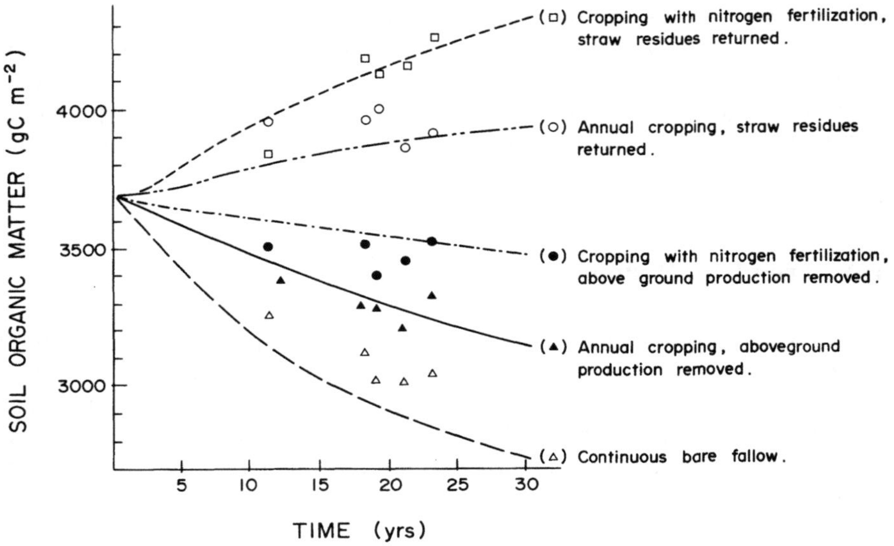

Figure 11. A comparison of simulation model predictions (lines) and actual measurements (symbols) for long-term agronomic experiments on Swedish soils. (From Parton et al., 1982). With permission of the International Society for Ecological Modelling.

This is the present challenge to those working in grassland soils: to understand the processes of the grassland ecosystem and managed equivalents and be able to predict the impact of future management alternatives. One small step in this direction is typified by output from a simulation model for organic matter formation as applied to soils under different managements (Parton et al., 1982). This simulation model was developed from a conceptual model resulting from pedological research (Anderson, 1979). Its application in assessments or predictions of soil management alternatives is demonstrated by the good agreement between simulations and the results of long-term field experiments (Figure 11).

XV. Summary

Pedogenic processes in the Great Plains grasslands are conditioned by factors as far distant in time and space as the orogeny that uplifted the western mountains and determined the major climatic and vegetation zones of the region. Locally, these regional effects are influenced by environmental conditions within hillslopes or catenas to determine the nature and intensity of the soil-forming processes.

Soils of semiarid grasslands have moderate equilibrium levels of organic matter, are not strongly weathered, and are well supplied with bases. With increasing moisture within the grasslands, greater biomass production results in higher organic matter contents and a shift to more strongly humified forms. There are concomitant increases in the degree of weathering and depth of leaching, resulting in more strongly developed soils that may be more acidic, and slightly lower in nutrients such as Ca, P, and S. Gradients to warmer climates result in lower organic matter equilibrium levels because of enhanced decomposition rates and increased weathering, other factors being equal.

Within landscapes, convex upland sites are locally more arid with consequent lower productivity and more weakly developed soils. Lower slope or concave positions are locally more moist, with deeper soils and higher organic matter contents.

Saline and solonetzic soils occur in physiographic areas where regional geohydrological conditions maintain high soluble salt or Na concentrations at or near the soil surfaces. Once again, local conditions, such as topography and near-surface stratigraphy, influence the balance between downward fluxes, which tend to move soluble components out of the pedosphere, and upward fluxes, which result form regional hydrological forces.

Soils formed on sediments rich in expanding clays, the Vertisols, have distinctive properties. They form wide cracks when dry, which provide channels for the entry of moisture and the displacement of surface material downward. Swelling due to moistening creates physical forces that churn the soil and mix it to considerable depth.

Moving to, and across, the grassland-forest ecotone, there is a marked increase in weathering and leaching, less below-ground additions of organic matter, and the buildup of surface litter horizons. Much enhanced clay translocation processes and stronger weathering environments in the upper solum result in the development of pronounced eluvial and illuvial horizons in forest soils where deep drainage is not impeded. Forest soils tend to be more acidic and lower in nutrients that are held in cationic form on the exchange complex, or associated with organic matter. Forests have a much higher proportion of their carbon and nutrients associated with the standing crop and litter layers than grassland soils.

Acknowledgments

The careful reviews of Dr. D.F. Acton and Sandra L. Anderson are gratefully acknowledged. The opportunity to examine soils in the Great Plains resulted from sabbatical leave at Colorado State University, and collaborative work with Drs. C.V. Cole and R.D. Heil.

References

Aandahl, A.R. 1948. The characterization of slope positions and their influence on the total *N* content of a few virgin soils of Western Iowa. *Soil Sci. Soc. Am. Proc.* 13:449–454.

Aandahl, A.R. 1982. *Soils of the Great Plains.* University of Nebraska Press., Lincoln, NB.

Allen, B.L., and D.S. Fanning. 1983. Composition and soil genesis. In: L.P. Wilding, N.E. Smeck, and G.F. Hall (eds.), *Pedogenesis and Soil Taxonomy: I. Concepts and Interactions.* Elsevier, New York, pp. 141–192.

Anderson, D.W. 1977. Early stages of soil formation on glacial till mine spoils in a semi-arid climate. *Geoderma* 19: 11–19.

Anderson, D.W. 1979. Processes of humus formation and transformation in soils of the Canadian Great Plains. *J. Soil Sci.* 30: 77–84.

Anderson, D.W. 1986. Topography and vegetation effects on soil development on a landscape in the aspen parkland. *Abstracts of the Annual Meeting of the Canadian Society of Soil Science, Saskatoon.*

Anderson, D.W., E. de Jong, and D.S. McDonald. 1979. The pedogenetic origin and characteristics of organic matter of Solod soils. *Can. J. Soil Sci.* 59: 357–362.

Anderson, D.W., and E.A. Paul. 1984. Organo-mineral complexes and their study by radiocarbon dating. *Soil Sci. Soc. Am. J.* 48: 298–301.

Anderson, D.W., R.E. Redmann, and M.E. Jonescu. 1975. *A Soil, Vegetation and Microclimate Inventory of Coal Strip-mine Wastes of* Estevan area, Saskatchewan. Saskatchewan Institute of Pedology, University of Saskatchewan, Saskatoon.

Anderson, D.W., D.B. Russell, R.J. St. Arnaud, and E.A. Paul. 1974. A comparison of humic fractions of Chernozemic and Luvisolic soils by elemental analyses, UV and ESR spectroscopy. *Can. J. Soil Sci.* 54: 447–456.

Anderson, D.W., S. Saggar, J.R. Bettany, and J.W.B. Stewart. 1981. Particle size fractions and their use in studies of soil organic matter: I. The nature and distribution of forms of carbon, nitrogen and sulfur. *Soil Sci. Soc. Am. J.* 45: 767–772.

Anderson, J.U., K.E. Fadul, and G.A. O'Connor. 1973. Factors affecting the coefficient of linear extensibility in Vertisols. *Soil Sci. Soc. Am. Proc.* 37: 296–299.

Arkley, R.J. 1967. Climates of some great soil groups of the western United States. *Soil Sci.* 103: 389–400.

Arshad, M.A., and R.J. St. Arnaud. 1980. Occurrence and characteristics of ferromanganiferous concretions in some Saskatchewan soils. *Can. J. Soil Sci.* 60: 685–695.

Baldwin, M., C.E. Kellogg, and J. Thorp. 1938. Soil classification. In: *Soils and Men. U.S.D.A. Yearbook of Agriculture*, pp. 979–1001, United States Government Printing Office, Washington D.C.

Bauer, A., and A.L. Black. 1981. Soil carbon, nitrogen and bulk density comparisons in two cropland tillage systems after 25 years and in virgin grassland. *Soil Sci. Soc. Am. J.* 45: 1166–1170.

Bauer, A., H.W. Omodt, and F.W. Schroer. 1979. *Effect of Nitrogen-Fertilized Wheat on Water Loss from Soils of the Barnes Catena.* (North Dakota Research Report No. 73). North Dakota State University, Fargo, North Dakota.

Baxter, F.P., and F.D. Hole. 1967. Ant (*Formica cinerea*) pedoturbation in a prairie soil. *Soil Sci. Soc. Am. Proc.* 31: 425–428.

Bentley, C.F., and C.A. Rost. 1947. A study of solonetzic soil complexes in Saskatchewan. *Sci. Agric.* 27: 293–313.

Bettany, J.R., H.H. Janzen, and J.W.B. Stewart. 1983. Sulfur deficiency in the prairie provinces of Canada. In: A.I. More (Ed.), *Proceedings of the International Sulphur '82 Conference.* British Sulphur Corp., London, Vol. 2, pp. 787–799.

Bettany, J.R., J.W.B. Stewart, and E.H. Halstead. 1973. Sulfur fractions and carbon, nitrogen and sulfur relationships in grassland, forest and associated transitional soils. *Soil Sci. Soc. Am. Proc.* 37:915–918.

Bettany, J.R., J.W.B. Stewart, and S. Saggar. 1979. The nature and forms of sulfur in organic matter fractions of soils selected along an environmental gradient. *Soil Sci. Soc. Am. J.* 43:981–985.

Bird, R.D. 1961. *Ecology of the Aspen Parkland.* Publication 1066, Canada Department of Agriculture, Ottawa.

Birkeland, P.W. 1984. *Soils and Geomorphology.* Oxford University Press, New York.

Brewer, R., and J.R. Sleeman. 1970. Some trends in pedology. *Earth Sci. Rev.* 6:297–335.

Buol, S.W., F.D. Hole, and R.J. McCracken. 1973. *Soil Genesis and Classification.* Iowa State University Press, Ames, Iowa.

Canada Soil Survey Committee (CSSC), Subcommittee on Soil Classification. 1978. *The Canadian System of Soil Classification* (Canadian Department of Agriculture Publication 1646). Supply and Services Canada, Ottwa, Ontario.

Coupland, R.T., and J.S. Rowe. 1969. The natural vegetation of Saskatchewan. In J.H. Richards and K.I. Fung (eds.) *Atlas of Saskatchewan.* University of Saskatchewan, Saskatoon, pp. 73–77.

Crocker, R.L. 1952. Soil genesis and the pedogenic factors. *Quart. Rev. Biol.* 27:139–168.

Dasog, G.S., D.F. Acton, and A.R. Mermut. 1986. Genesis and classification of clay soils in Saskatchewan, Canada. *Trans. XIII Congr. Int. Soc. Soil Sci.* III:1090–1091.

DeMent, J.A., and L.J. Bartelli. 1969. The role of vertic subgroups in the comprehensive soil classification system. *Soil Sci. Soc. Am. Proc.* 33:129–131.

de Sigmond, A.A. 1938. *The Principles of Soil Science.* Thomas Murby and Co., London.

Dijkerman, J.C. 1974. Pedology as a science: The role of data, models and theories in the study of natural ecosystems. *Geoderma* 11:73–93.

Dokuchaev, V.V. 1883. *Russian Chernozem.* Israel Program for Scientific Publication (1967).

Doughty, J.L., F.D. Cook, and F.G. Warder. 1954. Effects of cultivation on the organic matter and nitrogen of Brown soils. *Can. J. Agric. Sci.* 34:406–411.

Dudas, M.J., and S. Pawluk. 1969. Naturally occurring organo-clay complexes of Orthic Black Chernozems. *Geoderma* 3:5–17.

Ellis, J.G., D.F. Acton, and H.C. Moss. 1967. *The Soils of the Willowbunch Lake Map Area, 72H Saskatchewan.* Saskatchewan Institute of Pedology, University of Saskatchewan, Saskatoon.

Ellis, J.H. 1931. A field classification of soils for use in the soil survey. *Sci. Agric.* 12:338–345.

Ellis, J.H. 1938. *The Soils of Manitoba.* The Department of Agriculture and Immigration, Province of Manitoba, Winnipeg.

Fenster, C.R. 1977. Conservation tillage in the northern plains. *J. Soil Water Cons.* 32:37–42.

Fenton, T.E. 1983. Mollisols. In: L.P. Wilding, N.E. Smeck, and G.F. Hall (eds.), *Pedogenesis and Soil Taxonomy: II. The Soil Order.* Elsevier, New York, pp. 125–164.

Ferguson, H., and T. Bateridge. 1982. Salt status of glacial till soils of north-central Montana as affected by the crop-fallow system of dryland farming. *Soil Sci. Soc. Am. J.* 46:807–810.

Geis, J.W., W.R. Boggess, and J.D. Alexander. 1970. Early effects of forest vegetation and topographic position on dark-colored, prairie-derived soils. *Soil Sci. Soc. Am. Proc.* 34:105–111.

Graham, R.C., and A.R. Southard. 1983. Genesis of a Vertisol and an associated Mollisol in northern Utah. *Soil Sci. Soc. Am. J.* 47:552–559.

Gregorich, E.G., and D.W. Anderson. 1985. Effects of cultivation and erosion on the soils of four toposequences in the Canadian prairies. *Geoderma* 36:343–354.

Haas, H.J., C.E. Evans, and E.F. Miles. 1957. *Nitrogen and Carbon Changes in Great Plains Soils as Influenced by Cropping and Soil Treatments,* (U.S.D.A. Tech. Bull. 1164), U.S. Government Printing Office, Washington, D.C.

Henry, J.L., P.R. Bullock, T.J. Hogg, and L.D. Luba. 1985. Groundwater discharge from glacial and bedrock aquifers as a soil salinization factor in Saskatchewan. *Can. J. Soil Sci.* 65:749–768.

Hind, H.Y. 1859. *Narrative of the Canadian Red River Exploring Expedition of 1857, and of the Assiniboine and Saskatchewan Exploring Expedition of 1858.* Hurtig Publishers, Edmonton, Canada (Reprint).

Hole, F.D., and G.A. Nielsen. 1970. Soil genesis under prairie. In: P. Schramm (ed.), *Proceedings of the Symposiun on Prairie and Prairie Restoration.* Knox College, Galesberg, IL, Publ. 3, pp. 28–34.

Hoyt, P.B., W.A. Rice, and A.M.F. Hennig. 1978. Utilization of northern Canadian soils for agriculture. *Trans. 11th Int. Congr. Soil Sci.* 3:332–347.

Huang, P.M., and S.Y. Lee. 1969. Effect of drainage on weathering transformations of mineral colloids of some Canadian prairie soils. *Proceedings of the 3rd International Clay Conference.* Tokyo, Vol. 1, pp. 541–551.

Jackson, M.L. 1964. Chemical composition of soils. In F.E. Bear, (ed.), *Chemistry of the Soil* (ACS Monograph No. 160). Reinhold Publ. Corp., New York, pp. 71–134.

Jenny, H. 1941. *Factors of Soil Formation.* McGraw-Hill, New York.

Jenny, H. 1980. *The Soil Resources, Origin and Behavior.* Springer-Verlag, New York.

King, G.J., D.F. Acton, and R.J. St. Arnaud. 1983. Soil-landscape analysis in relation to soil distribution and mapping at a site within the Weyburn Association. *Can. J. Soil Sci.* 63:657–670.

Kittredge, J. 1948. *Forest Influences.* McGraw-Hill, New York.

Kodama, H. 1979. Clay minerals in Canadian soils: Their origin, distribution, and alteration. *Can. J. Soil Sci.* 59:37–58.

Kohnke, H., R.G. Stuff, and P.A. Miller. 1968. Quantitative relations between climate and soil formation. *Zeit. Pflanz. Boden.* 119:24–33.

Kolkhodzhayev, M.K. 1962. Amounts of plant residues and humus in some soils of southwestern Kalba and adjacent regions. *Sov. Soil Sci.*: 1217–1221.

Kononova, M.M. 1966. *Soil Organic Matter* (2nd Ed.). Pergamon Press, Oxford.

Kunze, G.N., H. Oakes, and M.E. Bloodworth. 1963. Grumusols of the coast prairie of Texas. *Soil Sci. Soc. Am. Proc.* 27:412–421.

Lauenroth, W.K., and W.C. Whitman. 1977. Dynamics of dry matter production in a mixed-grass prairie in western North Dakota. *Oecologia* 27:339–351.

Lewis, D.T. 1977. Subgroup designations of three Udolls in southeastern Nebraska. *Soil Sci. Soc. Am. J.* 41:940–945.

Lura, C.L., J.L. Richardson, and W.T. Barker. 1986. Soil organic matter and native plant production on the Missouri Coteau. *N. Dak. Farm Res.* 44-2:14–21.

MacLean, A.H., and S. Pawluk. 1975. Soil genesis in relation to groundwater and soil moisture near Vegreville, Alberta. *J. Soil Sci.* 26:278–293.

Malo, D.D., B.K. Worcester, D.K. Cassel, and K.D. Matzdorf. 1974. Soil landscape relationship in a closed drainage system. *Soil Sci. Soc. Am. Proc.* 38:813–818.

Maynard, D.G., J.W.B. Stewart, and J.R. Bettany. 1984. Sulfur cycling in grassland and parkland soils. *Biogeochemistry* 1:97–111.

McDaniel, P.A., and L.C. Munn. 1985. Effect of temperature on organic carbon-texture relationships in Mollisols and Aridisols. *Soil Sci. Soc. Am. J.* 49:1486–1489.

McKeague, J.A., M.V. Cheshire, F. Andreux, and J. Berthelin. 1986. Organo-mineral complexes in relation to pedogenesis. In P.M. Huang (ed.), *Mineral with Natural Organics and Microbes* (Soil Sci. Soc. Am. Spec. Publ. 17), pp. 549–592, Soil Science Soc. of America, Madison, Wisconsin.

Mermut, A.R., D.F. Acton, and W.D. Eilers. 1983. Estimation of soil erosion and deposition by a landscape analysis technique on clay soils in southwestern Saskatchewan. *Can. J. Soil Sci.* 63:727–739.

Mermut, A.R., and R.J. St. Arnaud. 1983. Micromorphology of some Chernozemic soils with grumic properties in Saskatchewan. *Soil Sci. Soc. Am. J.* 47:536–541.

Mitchell, J., H.C. Moss, and J.S. Clayton. 1944. *Soil Survey of Southern Saskatchewan* (Report No. 12). University of Saskatchewan, Saskatoon.

Moss, H.C., and J.S. Clayton. 1969. The soils of Saskatchewan. In: J.H. Richards and K.I. Fung (eds.), *Atlas of Saskatchewan.* University of Saskatchewan, Saskatoon, pp. 70–72.

Munn, L.C., and Boehm, M.M. 1983. Soil genesis in a Natrargid-Haplargid complex in northern Montana. *Soil Sci. Soc. Am. J.* 47:1186–1192.

Munn, L.C., G.A. Nielsen, and W.F. Mueggler. 1978. Relationships of soils to mountain and foothill range habitat types and production in western Montana. *Soil Sci. Soc. Am. J.* 42:135–139.

Newton, J.D., F.A. Wyatt, and A.L. Brown. 1945. Effects of cultivation and cropping on the chemical composition of some western Canada Prairie Province soils, Part III. *Sci. Agric.* 26:718–738.

Nichols, J.D. 1984. Relation of organic carbon to soil properties and climate in the southern Great Plains. *Soil Sci. Soc. Am. J.* 48:1382–1384.

Nikiforoff, C.C. 1938. Soil organic matter and soil humus. In: *Soils and Men. U.S.D.A. Yearbook of Agriculture*, pp. 929–939, United States Government Printing Office, Washington, D.C.

Nikiforoff, C.C. 1959. Reappraisal of the soil. *Science* 129:186–196.

Nyhan, J.W. 1976. Influence of soil temperature and water tension on the decomposition rate of carbon-14 labeled herbage. *Soil Sci.* 121:288–293.

Parton, W.J., J. Persson, and D.W. Anderson. 1982. Simulation of organic matter changes in Swedish soils. In: W.K. Lauenroth, G.V. Skogerboe, and M. Flug (eds.), *Analysis of Ecological Systems: State-of-the-Art in Ecological Modelling (Developments in Environmental Modelling 5)*. Elsevier Science, Amsterdam, pp. 511–516.

Pawluk, S. 1982. *Salinization and Solonetz Formation.* Proceedings of the 19th Alberta Soil Science Workshop, Edmonton, Canada.

Pawluk, S. 1985. Soil micromorphology and soil fauna, problems and importance. *Quaestiones Entomologicae* 21:473–496.

Pettapiece, W.W. 1969. The forest-grassland transition. In: S. Pawluk, (ed.), *Pedology and Quaternary Research.* University of Alberta, Edmonton, pp. 103–113.

Rennie, D.A. 1982. The deteriorating soils of the Canadian Prairie. *Span* 25:3.

Rennie, D.A., and J.S. Clayton. 1967. An evaluation of techniques used to characterize the comparative productivity of soil properties in Saskatchewan. *Trans. Comm. II and IV*, ISSS, Aberdeen, pp. 365–376.

Richards, J.H. 1969. Physical features of Saskatchewan. In: J.H. Richards, and K.I. Fung (eds.), *Atlas of Saskatchewan*. University of Saskatchewan, Saskatoon, pp. 40–41.

Risser, P.G., E.C. Birney, H.D. Blocker, S.W. May, W.J. Parton, and J.A. Wiens. 1981. *The True Prairie Ecosystem*. Hutchinson Ross Publishing Co., Stroudsburg, PA.

Roberts, T.L., and J.R. Bettany. 1985. The influence of topography on the nature and distribution of soil sulfur across a narrow environmental gradient. *Can. J. Soil Sci.* 65:419–434.

Roberts, T.L., J.R. Bettany, and J.W.B. Stewart. 1986. The level-of-integration approach to the study of organic C, N, P and S in the Canadian Prairies. *Trans. XIII Congr. Int. Soc. Soil Sci.* III:932–933.

Rowe, J.S. 1961. The level-of-integration concept and ecology. *Ecology* 42:420–427.

Sanborn, P., and S. Pawluk. 1983. Process studies of a Chernozemic pedon, Alberta (Canada). *Geoderma* 31:205–237.

Schimel, D.S., D.C. Coleman, and K.A. Horton. 1985. Soil organic matter dynamics in paired rangeland and cropland toposequences in North Dakota. *Geoderma* 36: 201–214.

Schimel, D.S., M.A. Stillwell, and R.G. Woodmansee. 1985. Biogeochemistry of C, N, and P in a soil catena of the shortgrass steppe. *Ecology* 66:276–282.

Schimel, D.S., W.J. Parton, F.J. Adamson, R.G. Woodmansee, R.L. Senft, and M.A. Stillwell. 1986. The role of cattle in the volatile loss of nitrogen from a shortgrass steppe. *Biogeochemistry* 2:39–52.

Schoenau, J.J., and J.R. Bettany. 1987. Organic matter leaching as a component of C, N, P and S cycles in a forest, grassland and gleyed soil *Soil Sci. Soc. Am. J.* 51:646–651.

Severson, R.C., and H.F. Arneman. 1973. Soil characteristics of the forest-prairie ecotone in northwestern Minnesota. *Soil Sci. Soc. Am. Proc.* 37:593–599.

Simonson, R.W. 1959. Outline of a generalized theory of soil genesis. *Soil Sci. Soc. Am. Proc.* 23:152–156.

Sims, P.L., and J.S. Singh. 1978. The structure and function of ten western North American grasslands: III. Net energy production, turnover and efficiencies of energy capture and water use. *J. Ecol.* 66:573–597.

Sims, Z.R., and G.A. Nielsen. 1986. Organic C in Montana soils as related to clay content and climate. *Soil Sci. Soc. Am. J.* 50:1269–1271.

Smeck, N.E. 1973. Phosphorus: an indicator of pedogenic weathering processes. *Soil Sci.* 115:199–206.

Smeck, N.E., and E.C.A. Runge. 1971. Phosphorus availability and redistribution in relation to profile development in an Illinois landscape segment. *Soil Sci. Soc. Am. Proc.* 35:952–959.

Smeck, N.E., E.C.A. Runge, and E.E. Mackintosh. 1983. Dynamics and modelling of soil systems. In: L.R. Wilding, N.E. Smeck, and G.F. Hall (eds.), *Pedogenesis and Soil Taxonomy: I. Concepts and Interactions*. Elsevier, New York, pp. 51–79.

Soil Survey Staff. 1975. *Soil Taxonomy*, (Agric. Handbook No. 436). USDA-SCS, U.S. Government Printing Office, Washington, DC.

Somasiri, S., and P.M. Huang. 1971. The nature of K-feldspars of a Chernozemic soil in the Canadian prairies. *Soil Sci. Soc. Am. Proc.* 35:810–815.

Southard, A.R., and Dirmhirn. 1972. Some observations on soils—microclimate interactions. *Soil Sci. Soc. Am. Proc.* 36:843–845.

Spiers, G.A., S. Pawluk, and M.J. Dudas. 1984. Authigenic mineral formation by solodization. *Can. J. Soil Sci.* 64:515–532.

St. Arnaud, R.J. 1979. Nature and distribution of secondary soil carbonates within landscapes in relation to soluble Mg^{++}/Ca^{++} ratios. *Can. J. Soil Sci.* 59:87–98.

St. Arnaud, R.J. 1986. The "Pedogenic Index"—its application to mineral weathering and soil fertility. *Trans. XIII Congr. Int. Soil Sci. Soc.* III:1042–1043.

St. Arnaud, R.J., and M.M. Mortland. 1963. Characteristics of the clay fractions in a Chernozemic to Podzolic sequence of soil profiles in Saskatchewan. *Can. J. Soil Sci.* 43:336–349.

St. Arnaud, R. J., and M.D. Sudom. 1981. Mineral distribution and weathering in Chernozemic and Luvisolic soils from central Saskatchewan. *Can. J. Soil Sci.* 61:79–89.

St. Arnaud, R.J., and E.P. Whiteside. 1964. Morphology and genesis of a Chernozemic to Podzolic sequence of soil profiles in Saskatchewan. *Can. J. Soil Sci.* 44:88–99.

Stewart, J.W.B., C.V. Cole, and D.G. Maynard. 1983. Interaction of biogeochemical cycles in grassland ecosystems. In: B. Bolin and R.B. Cook, (eds.) *The Major Biogeochemical Elements and Their Cycles.* Wiley, New York, pp. 247–269.

Thorp, J. 1947. Practical problems in soil taxonomy and soil mapping in Great Plains States. *Soil Sci. Soc. Am. Proc.* 12:444–448.

Thorp, J., B.H. Williams, and W.I. Watkins. 1948. Soil zones of the Great Plains states, Kansas to Canada. *Soil Sci. Soc. Am. Proc.* 13:438–445.

Tiessen, H., J.W.B. Stewart, and J.R. Bettany. 1982. Cultivation effects on the amounts and concentration of carbon, nitrogen and phosphorus in grassland soils. *Agron. J.* 74:831–835.

Tiessen, H., J.W.B. Stewart, and C.V. Cole. 1984. Pathways of phosphorus, transformations in soils of differing pedogenesis. *Soil Sci. Soc. Am. J.* 48:853–858.

Trudgill, S. 1977. *Soil and Vegetation Systems.* Oxford University Press, Oxford.

Tyurin, I.V., I.P. Gevasimov, E.N. Ivanova, and V.A. Nosin (eds.). 1959. *Soil Survey, a Guide to Field Investigations and Mapping of Soils.* Israel Program for Scientific Translations Ltd., Jerusalem.

Volkovintser, V.I. 1969. Soil formation in the steppe basins of southern Siberia. *Sov. Soil Sci.* 4:383–391.

Volobuev, V.R. 1964. *Ecology of Soils.* Israel Program for Scientific Translations Ltd., Jerusalem.

Voroney, R.P., J.A. Van Veen, and E.A. Paul. 1981. Organic C dynamics in grassland soils: 2. Model validation and simulation of long-term effects of cultivation and rainfall erosion. *Can. J. Soil Sci.* 61:211–224.

Warkentin, J. 1969. *The Western Interior of Canada.* McClelland and Stewart Ltd., Toronto, Canada.

Westin, F.C. 1976. Geography of soil test results. *Soil Sci. Soc. Am. J.* 40:890–895.

Westin, F.C., and G.J. Buntley. 1967. Soil phosphorus in South Dakota: III. Phosphorus fractions of some Borolls and Ustolls. *Soil Sci. Soc. Am. Proc.* 31:521–528.

White, E.M. 1964. Morphological chemical problems in solodized soil. Soil Sci. 98:187–189.

Wildung, R.E., T.R. Garland, and R.L. Buschbom. 1975. The interdependent effects of soil temperature and water content on soil respiration rate and plant root decomposition in arid grassland soils. *Soil Biol. Biochem.* 7:373–378.

Woodmansee, R.G., I. Vallis, and J.J. Mott. 1981. Grassland nitrogen. In: F.E. Clark and T. Rosswall (ed.), *Terrestrial Nitrogen Cycles. Ecol. Bull.* (Stockholm) 33:443–462.

Legume Winter Cover Crops

M. Scott Smith, Wilbur W. Frye, and Jac J. Varco*

*Department of Agronomy, University of Kentucky, Lexington, Kentucky 40546, U.S.A.

I. Introduction

Throughout virtually all of the history of agriculture, the nitrogen harvested from cropped soils has been replenished, if it has been replenished at all, by leguminous nitrogen fixation. Although animal wastes, nonsymbiotic fixation, and atmospheric deposition can be significant sources of N, a large fraction of the first can be traced to legume sources and the latter two are generally insufficient to maintain productivity of cropland. In the Mediterranean Civilizations, documented recognition of the value of green manures can be found as early as the writings of Xenophon, who lived from 434 to 355 B.C. (according to Wedderbuan and Collingwood, 1976). Semple (1928), in a review of ancient agricultural practices, indicated that several writers have specifically discussed the use of legumes for soil improvement. Theophrastus (373–287 B.C.) wrote of bean crops being used as green manure by farmers of Macedonia and Thessaly. Cato (234–149 B.C.) and Columella (about 45 A.D.) compared the value of various legumes in soil improvement. Lupine was a favored legume for this purpose. According to Pieters (1927), Chinese writers recognized more than 2,000 years ago that legumes increased production of the crops that followed. As is often the case now, development of these practices by farmers may considerably predate their consideration by academics.

At the very beginning of modern agricultural science, Lawes and Gilbert conducted experiments to measure and understand the contributions of legumes to soil fertility (Russell, 1966). By the 1930s, the mechanisms by which legumes enhance soil N availability, N_2 fixation, and organic N mineralization, had become reasonably well understood (Waksman and Starkey, 1931; Fred et al., 1932).

In early American agriculture it is difficult to determine how widely green manuring and cover cropping were practiced. According to Pieters and McKee (1929), these practices were known but not common in colonial times. As native soil fertility was depleted, the value of legume manures should have become more apparent, but perhaps this problem was more commonly solved by long-term pasture rotations, application of animal manure to grain crops, or moving west. The era from 1900 to World War II was a very active period of research and extension activity on legumes for soil improvement (e.g., Lohnis, 1926; Pieters, 1927; Lyon, 1936; Coleman, 1941). A.J. Pieters was perhaps most active in spreading this gospel. For good reasons, to be considered below, much of this activity occurred in the Southeast and Mid-Atlantic regions. But even here, it is not clear that legume cover crops were in common use. In 1936, which other authors have suggested is near the heyday of green manuring, there were 55 million hectares of cropped land in 12 southern states but only 0.8 million were seeded to winter cover crops (Pieters and McKee, 1938). Rogers and Giddens (1957) reported that there were 5.9 million hectares of green manure crops in the Southeast in 1940 but that acreage declined significantly after this time. This decline can be attributed to the widespread availability of synthetic N fertilizer and the economic advantage of its use in continuous grain crop production systems.

After World War II, agronomists did not completely ignore winter cover crops but, for the most part, farmers did. Several excellent studies were conducted between 1940 and 1965 (Nelson, 1944; Evans et al., 1954; Beale et al., 1955; Kamprath et al., 1958; Benoit et al., 1962; and others), but the practice was rather limited.

Considering this history, it would certainly be difficult to claim that legume cover crops are a new idea. However, during the last 10 years there has clearly been a tremendous renewal of interest, on the part of both researchers and producers, in this old practice. This can be related to three major factors. First were the large increases in the cost of fossil fuels and the related increases in the price of N fertilizer experienced in the 1970s and early 1980s. Although costs of both commodities have stabilized or even decreased more recently, the perception remains that over the long-term these are likely to become more expensive or more limited in supply. A second factor is the increased concern about soil erosion and more general concern about the effects of agricultural practices on environmental quality. Third is the rapid adoption of no-tillage and conservation tillage by crop producers in many regions of the United States and throughout the world. Most of the recent work on legume cover crops has been done in conservation tillage systems. It has been seen that legume cover crops fit well and particularly in no-tillage.

This chapter reviews some new research on a very old idea. The emphasis is on legume winter cover crops in minimum-tillage grain crop systems, because this is their most promising application and because this is the focus of most current research. Consideration is given to what has been learned (and to some extent relearned) and what is not known about the agronomic advantages and disadvantages of legume cover crops, their effects on physical and chemical behavior of soils, and nitrogen cycling in these soils.

II. Production of Legume Winter Cover Crops

A. Suitability by Location

It should be possible to grow winter annual legumes throughout much of the temperate and subtropical regions of the world. Yet the locations where reasonable production can be expected and where they are likely to be beneficial as a winter cover for summer crops are considerably more restricted. These restrictions are imposed by temperature and water availability. In the United States, the southeastern quarter has historically been the area of greatest use of legumes for winter cover crops. The current revival of interest is also centered there. Therefore, this review has a distinct southern accent. Although the authors are aware that some research has begun outside of the mid-Atlantic coast and Southeast, results are not available at present.

Temperature limitations are likely to be severe north of southern Ohio and Pennsylvania. Frequent winterkill of fall-seeded legumes is probable, and even if there is adequate survival, it is unlikely that there will be enough warm days in early spring for good legume growth. This problem might be overcome by delayed planting of the summer crop, but in the North this will almost always cause significant yield reductions and will rarely be an economical option. In 9 years of experimentation with hairy vetch (*Vicia villosa*) winter covers in central Kentucky, we have experienced only 1 year of severe winterkill. However, other winter annual legumes have suffered severe over-winter losses.

Water availability is more likely to present problems in the Southwest and southern plains. Cover crops will deplete soil moisture reserves on which the cash crops are dependent. On irrigated lands it is doubtful that the production of legumes for use solely as green manure currently would be profitable. On the other hand, the accelerated loss of soil water in the spring is less likely to be detrimental in the Southeast and sometimes, particularly on poorly drained soils, will be advantageous.

A third factor making the Southeast more likely to find advantages in legume winter covers is the nature of the soils. Relatively weathered, low organic matter soils are characteristic of much of this region. Continuous row cropping of such soils generally will result in faster and greater losses in productivity than will be experienced on Mollisols of the Great Plains, for example. Agronomists and many farmers of the Southeast learned long ago that legume covers offered one way of sustaining soil productivity.

Although most of the current research on legume winter cover crops is being done in the Southeast, it is expected that these systems may be applicable to other parts of the world with similar climates and soils. Also, the knowledge acquired in the development of these practices should be helpful in designing new cropping systems that utilize legumes as a N source and mulch for orchards, vegetables, or in rotations with grain and fiber crops.

B. Suitability for Minimum Tillage

Perhaps the most significant recent change in American agriculture is the rapid adoption of conservation tillage practices (Phillips and Phillips, 1984; D'Itri, 1985). Conservation tillage has become common because it can: save the farmer time, labor, tractor fuel, and machinery costs; minimize soil erosion; conserve soil water; maintain or sometimes increase yields; and often increase profits. Minimum and conservation tillage have been given a variety of definitions and these practices have taken many forms. They have in common the abandonment of the moldboard plow and the retention of much of the crop residue on the soil surface. No-tillage, which is an extreme (some would say ultimate) version of minimum tillage, has been used in most of the recent studies with legume cover crops. In no-tillage, the soil is undisturbed except for a narrow slit required for seed placement.

Winter legume cover crops are particularly well suited where continuous no-tillage corn (*Zea mays*) or grain sorghum (*Sorghum bicolor*) is grown, because including legumes in this system requires few management changes. Many farmers using no-tillage techniques are already using a grass cover crop, most commonly rye (*Secale cereale*) (Blevins et al., 1977). These were used for further control of soil erosion and to provide additional moisture-conserving mulch. With no-tillage, good cover crop establishment can be achieved in early autumn simply by broadcasting seed into the standing summer crop. The cover crop seed germinate under the mulch on the moist soil surface. In the lower South, where corn may be harvested as early as August, the cover crop can be seeded after harvest with a no-tillage planter. The time available for cover crop growth is maximized, since soil tillage is not required to establish a seedbed for the summer crop. A vigorous cover crop can also suppress the growth of weeds, which are more difficult to control in no-tillage (Witt, 1984). It was logical to seek these advantages plus the additional benefit of supplemental N input by using winter annual legumes in place of grasses as a cover crop.

C. Management Practices

Management practices for legume cover crops in Kentucky experiments (Ebelhar et al., 1984) are described briefly because they are similar to practices used elsewhere and will give an indication of how the system operates. In mid-September, inoculated legume seed are broadcast overseeded into standing corn. This is done by hand on experimental plots, but aerial seeding would be practical on farmers' fields. By the time corn is harvested in early October, legume seedlings have emerged. Some growth occurs in the fall, but much more dry matter accumulates in the spring. Most legumes are in bloom and have not set seed when corn is planted with a no-tillage planter in mid-May (Figure 1). Weeds and the cover crop can often be adequately controlled with

Figure 1. No-tillage corn planting into hairy vetch winter cover crop.

a single herbicide application at planting. Mixtures of paraquat and a triazine herbicide are commonly used. Additional postemergence herbicide applications have been used in some experiments (e.g., Touchton et al., 1982). Fertilizers, as required, are broadcast on the soil surface.

Legumes may be drilled in after the summer crop is harvested rather than overseeded. This will be advantageous if the soil is dry, however, it will delay legume establishment by approximately 1 month. This delay can significantly reduce cover crop growth as observed by Brown et al. (1985). Hairy vetch and bigflower vetch (*Vicia grandifiora*) generally produced more when overseeded in mid-September than when drilled in mid-October in a study by Varco et al. (1984).

In the lower South, where the growing season is longer, delays in legume seeding tend not to be a serious constraint and there is more flexibility in scheduling operations. In Georgia, it is generally recommended that legumes be planted about 1 month prior to the first frost date (W.L. Hargrove, personal communication). There may also be an advantage to killing the cover crop prior to planting the summer crop when there is a sufficiently long growing season. As discussed later, longer seasons also offer the opportunity to manage legumes for reseeding.

D. Productivity of Various Winter Legumes

Some progress apparently has been made in the identification and selection of highly productive winter annual legumes. However, the minimal effort that has been made in genetic improvement and evaluation suggests there is considerable opportunity for improvement. Many species have been used as winter cover crops. The most common ones in the Southeast are listed in Table 1, with some data on dry matter yields.

Crimson clover (*Trifolium incarnatum*) has been the most widely successful cover crop in recent years, with dry matter yields commonly about 5 Mg

Table 1. Comparison of total aboveground dry matter accumulation in different legume cover crops from several studies

	Dry matter yield (Mg ha^{-1}) from reference[a]						
Cover crop	1	2	3	4	5	6	7
Crimson clover	2.4	4.5	5.0	5.3	6.7	—	3.2
Hairy vetch	4.2	—	—	3.8	4.2	2.7	2.7
Common vetch	—	5.0	3.7	—	4.3	—	—
Bigflower vetch	2.2	—	—	—	—	—	—
Subterranean clover	—	—	3.8	6.0	4.0	—	—
Winter peas	—	—	—	5.6	—	1.7	1.8

[a] References: (1) Ebelhar et al. (1984) and Utomo (1986), Kentucky; (2) Touchton et al. (1984), Alabama; (3) Dabney et al. (1984), Louisiana; (4) Hoyt and Hargrove (1986), North Carolina; (5) Hargrove (1986), Georgia; (6) Kamprath et al. (1958), North Carolina; (7) Welch et al. (1950), North Carolina.

ha^{-1}. The superior performance of crimson clover in the deep South was extensively documented in an Alabama study (Donnelly and Cope, 1961). Four winter legumes were compared for the impressive total of 106 site-years. Yields of 11,100, 9,300, 10,400, and 6,800 were reported as fresh weight per acre for crimson clover, hairy vetch, Monantha vetch, and Austrian winter peas (*Pisum sativum*), respectively. Crimson clover has very good fall and early season growth and matures earlier than the vetches (Hoyt and Hargrove, 1986). The latter characteristic offers the possibility of self-reseeding in warmer climates, which might be a highly significant advantage as discussed later. It is moderately tolerant of soil acidity (Palada et al., 1983). Some improved cultivars are available. Crimson clover is clearly less cold tolerant than some other winter annual legumes and in Kentucky has consistently been outperformed by hairy vetch (Table 1).

Hairy vetch, being among the most winterhardy of legume cover crops, appears to have the greatest potential in the southern Midwest and mid-Atlantic regions. It has been very productive in Kentucky (Ebelhar et al., 1984) and Delaware (Mitchell and Teel, 1977), but it has long been one of the most common cover crops throughout the Southeast and is well adapted to a wide area. Common vetch (*Vicia sativa*) is less cold tolerant, but some improved varieties may be more productive than hairy vetch in the deep South (Touchton et al., 1984).

Several species of *Trifolium* other than crimson clover have been used as winter covers. Subterranean clover (*Trifolium subterraneum*) has generally been less productive than crimson clover, except in a recent North Carolina study (Hoyt and Hargrove, 1986). Red clover (*Trifolium pratense*) has been very extensively employed as a green manure crop throughout the country, but its generally greater sensitivity to stress and low management limits its role as a winter annual. Hoyt and Hargrove (1986) have reviewed the characteristics of many other species of legume covers. Among the other legumes that have been evaluated are Austrian winter pea, many more *Trifolium* species, some lupines, cowpeas (*Vigna sinensis*), and a few *Medicago* species.

Table 1 includes, for comparison, data from two reports in the 1950s. We believe these to be representative of legume cover crop yields in experiments of that era. It can be seen that these values are generally well below those reported more recently. This apparent improvement might be due to the greater legume growth period permitted in recent no-tillage studies or to some genetic improvement. Most older studies do not specify varieties, but it is assumed that unimproved lines and local seed sources were generally used. Identification of productive, well-adapted existing genotypes and development of new genotypes specifically suited for winter cover cropping should be a high priority in research on legume winter covers. A current cooperative project, which will test a number of legumes in several southeastern states, is a promising first step (W.L. Hargrove and R.L. Blevins, personal communication). An ideal cultivar for winter covers would be cold tolerant, early maturing, and highly productive. However, it may be that these characteristics are to some extent physiologically opposed (W.E. Knight, personal communication).

E. Nitrogen Accumulation

Table 2 shows aboveground N accumulation in several common winter legumes and in rye covers for five recent studies. Values in excess of 100 kg ha^{-1} are common for legumes, while rye generally accumulates less than 50. Very few investigators have made the necessarily tedious measurements of dry matter or N content in roots. Data from Mitchell and Teel (1977) suggest that the root contribution to total N will be small in winter annuals (Table 3). Less than one third of the total dry matter was found below ground and the N concentration in roots was no more than half that in tops. The percentage of total N accumulated that was in roots ranged from 8 to 23 and averaged 12% for all seven cover crops studied by Mitchell and Teel (1977).

Dry matter yield alone may be less important than the N input from legume cover crops. It is clear that the relationship between these two parameters is variable. As shown in Table 4, N concentration in winter legumes may vary more among different species than among different experiments. It would be surprising if experimental conditions had no effect on N concentration, yet

Table 2. Comparisons of aboveground nitrogen accumulation in legume and rye winter cover crops and estimation of the fraction of legume N derived from N_2 fixation by difference

Reference and location	Cover crop	N accumulated (kg ha^{-1})	Estimate of N from fixation
Brown et al. (1985), Alabama	Hairy vetch	133	0.79
	Crimson clover	133	0.79
	Rye	28	—
Varco et al. (1984), Kentucky	Hairy vetch	132	0.84
	Bigflower vetch	84	0.75
	Rye	21	—
Ebelhar et al. (1984), Kentucky	Hairy vetch	209	0.83
	Bigflower vetch	60	0.40
	Crimson clover	56	0.36
	Rye	36	—
Hargrove (1986), Georgia	Hairy vetch	153	0.75
	Crimson clover	170	0.78
	Subterranean clover	114	0.67
	Common vetch	134	0.72
	Rye	38	—
Mitchell and Teel (1977), Delaware	Hairy vetch + rye	158	0.71
	Crimson clover + rye	147	0.69
	Rye alone	46	—

Table 3. Nitrogen in roots and tops for different winter cover crops

Cover crop	N ($kg\ ha^{-1}$) accumulated in		N concentration ($g\ kg^{-1}$) in		Fraction of total N in roots
	Roots	Tops	Roots	Tops	
Spring oats	4	46	6	13	0.08
Rye	14	46	5	10	0.23
Rye + crimson clover	20	147	9	18	0.12
Rye + hairy vetch	16	158	7	20	0.09
Ryegrass + crimson clover	28	182	7	22	0.13

Adapted from Mitchell and Teel (1977)

Table 4. Nitrogen concentration in aboveground portion of winter legume cover crops

Cover crop	Nitrogen concentration ($g\ kg^{-1}$) from reference[a]			
	1	2	4	5
Crimson clover	24	22	23	25
Hairy vetch	41	—	39	36
Common vetch	—	24	—	31
Subterranean clover	—	—	27	28

[a] See footnote in Table 1 for references.

reasonably consistent results have been reported. Hairy vetch generally has been observed to contain at least 1% more N than crimson or other clovers. This might be explained by the earlier maturation of some clovers relative to hairy vetch. Or this may reflect the greater content of structural material in the more upright clovers versus the viny, thin-stemmed hairy vetch. The result is that hairy vetch may yield less dry matter than crimson clover but accumulate comparable amounts of N (Hoyt and Hargrove, 1986). When various legume covers are evaluated only on the basis of kilograms of nitrogen accumulated per hectare, hairy vetch is consistently as good or better than other species (Table 2).

This observation suggests that it may not always be appropriate to select for early maturation of cover crops to obtain reseeding and maximum dry matter production alone. The rate of N_2 fixation by many annual legumes is known to decline during maturation (Herridge and Pate, 1977). Hardy et al. (1973) have reported that early flowering soybean (*Glycine max*) cultivars fixed less total N than later maturity groups. It may be that N_2 fixation rates in crimson clover decline before the cover crop is killed and at a time when hairy vetch activity is maximal. The price for early maturation may be a relatively reduced potential for N_2 fixation and less N available to the summer crop.

There have been no direct measurements of the fraction of cover crop N that is derived from atmospheric N_2 fixation versus that derived from soil N. To understand N cycling in these systems, it is important to know not just total N accumulation, but also the quantity of additional N inputs from fixation. In Table 2, an indirect method of estimating fixation is used on data from five locations. If it is assumed that the legume cover and a nonlegume cover, rye in these cases, take up the same quantity of soil N and that an insignificant amount of the rye N is derived from fixation, then legume fixation can be estimated simply by the difference in N accumulated by the grass and the legume. The first of these assumptions is admittedly questionable. The estimated fractions of legume N derived from fixation are as low as 0.36 and 0.40 for an experiment in Western Kentucky where poor growth of crimson clover and bigflower vetch were observed. Where there was good legume growth, much more of the N was apparently derived from fixation, values ranging from 0.67 to 0.84.

Better methods of measuring long-term N_2 fixation include the use of a non-nodulating isoline of the same legume as a control, and the isotope dilution technique. The former has been used almost exclusively on soybean, which is the only legume for which nonnodulating genotypes are readily available. A nonnodulating genotype of crimson clover has recently been described by Smith and Knight (1984a). Although not yet available, this will soon be of great value in studies of nitrogen fixation. The absolute validity of the isotope dilution technique (reviewed by Chalk, 1985; Danso, 1986) is not universally accepted, but this is probably the most useful method available. In this approach, it is necessary to assume only that the isotopic composition of N taken up from ^{15}N-labeled soil is the same for a legume and a nonlegume control; it is not necessary that the amount of N taken up be the same. In preliminary experiments in Kentucky, fall-planted tall fescue is being used as the control because the temporal pattern of its growth best matches the winter legumes.

Over the last decade considerable research has been directed toward enhancing biological N_2 fixation in legumes. This has involved the selection of superior plant and rhizobium genotypes, including the consideration of plant-microbe interactions, refinement of inoculation technology and increased understanding of rhizobium ecology, and improved management practices (Smith and Knight, 1984b). Most of this effort has been focused on legumes for grain or forage production; fewer studies have been done with common winter cover crop legumes. However, several publications by W.E. Knight, C. Hagedorn, and collaborators, with crimson clover, clearly indicate the value of these approaches (Materon and Hagedorn, 1982; Smith, Knight, and Peterson, 1982; Smith, Knight, Peterson et al., 1982).

III. Yield Responses of Summer Crops

At low N rates, increased yields of crops following legumes, in rotations or as cover crops, have been well documented for many years at many locations. A full review of this particular point will not be presented here, but many supporting references are provided throughout this chapter and N effects of legumes are discussed specifically below. At this point, we would like to consider the following questions: Are yield responses after winter legumes attributable exclusively to N availability? What are the mechanisms and the relative significance of other stimulatory or inhibitory effects?

Figure 2 shows three hypothetical cases where yield of the summer crop as a function of N fertilizer applied is compared for legume cover crops versus a nonlegume or no cover crop. In the first case (A in Figure 2), there is a yield benefit from the legume at low N rates but no effect at high N rates. Such a response pattern would suggest that the only significant effect of the legume is to increase N supply to the summer crop. The results of Hargrove (1986), who compared four different legume covers with winter fallow or rye covers for no-tillage sorghum, fit this pattern. Flannery (1981) also illustrates this situation for rye compared to hairy vetch covers for no-tillage corn silage, as do Kamprath et al. (1958) and Welch et al. (1950) in older work with conventional tillage systems.

In the second case (B in Figure 2), yields with the legume cover crop are greater even at high N rates. The response to legumes may be greater at low N fertilizer rates, but in this case there is an indication that the legume provides benefits beyond N supply. Evidence for this pattern can be found in work by Moschler et al. (1967), Ebelhar et al. (1984), Touchton et al. (1984), and Mitchell and Teel (1977) for no-tillage cover cropping experiments and Baldock et al. (1981), Kumar Rao et al. (1983), and Yogeswara Rao and Ali (1983) for rotation or conventional tillage experiments. It should be emphasized that additional legume-derived benefits beyond N were not observed consistently (only in some years) in some of these studies (e.g., Touchton et al., 1984). In others, it was not certain that N supply was sufficient even at the highest rates

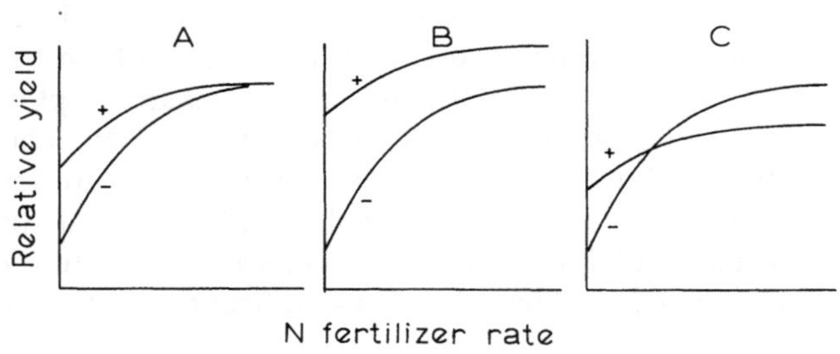

Figure 2. Hypothetical yield curves as a function of N fertilizer rate for summer crops following winter legumes (+) or without legume covers (−).

of applied fertilizer; that is, cover crop responses might have disappeared if more N fertilizer was used (e.g., Ebelhar et al., 1984).

Several possible explanations for this response pattern are discussed later in this review. These include more favorable moisture, temperature or soil physical conditions, or, in the long term, reduced erosion loss. Additional mechanisms are only speculative since reliable data are scarce. These include effects of cover cropping on insect or microbial pests or, conceivably, growth-regulating chemicals. Nonnutritional responses to legume cover crops are like the so-called rotation effect in that they are much discussed, widely accepted as real, difficult to reproduce consistently, and largely undefined.

In evaluating the relative significance of N-related versus non-N-related effects of legume cover crops, the most definite general statement that can be made is that the N contribution is almost always observed but that additional benefits are not. Attempts to partition out N supply from other contributions (e.g., Baldock et al., 1981) will always be site and year specific. Documentation of more favorable soil conditions will not establish that these factors necessarily contribute to better crop growth, as illustrated by the work of Wade and Sanchez (1983).

In the third hypothetical case (C in Figure 2), yields following legume cover crops are lower than those after no cover crop when sufficient N fertilizer is applied. There may still be a N contribution from the legume cover, but there are also negative effects associated with the legume. This was seen in some of the treatment comparisons by Brown et al. (1985), who observed poor cotton (*Gossypium hirsutum*) seedling performance and reduced stands with some legume systems. Worsham (1986) also discusses some evidence of inhibitory effects on corn and cotton with crimson clover cover crops. In preliminary work, seedling inhibition by water extracts of crimson clover was observed, raising the specter of allelopathy.

Negative responses to hairy vetch were also observed in Kentucky (S. Corak, W. Frye, and M. Smith, unpublished data). This effect on no-tillage corn yields and its interaction with N fertilizer rate are shown in Figure 3A. It is interesting that this was observed on a very similar soil and within 300 m of the experiment described by Ebelhar et al. (1984), where, in the same year, positive responses to vetch were observed with N fertilizer applications up to 150 kg ha^{-1}. The negative effects of vetch were minimal when cover crop top growth was removed (compare no vetch with 225 kg N in Figure 3A to roots only with 225 kg N in Figure 3B) and were increased on plots receiving double the normal vetch rate (with 225 kg N in Figure 3B). Therefore, the effect is due to the residues and not to early season moisture loss. Neither could the yield reduction be adequately explained by reduced stands in the heavy mulch. Water availability was excellent, so the moisture-conserving benefits of the mulch would not have been as significant as they would in many seasons.

The possible importance of allelopathic chemicals has generated much interest, particularly in minimum tillage systems. For example, it has long

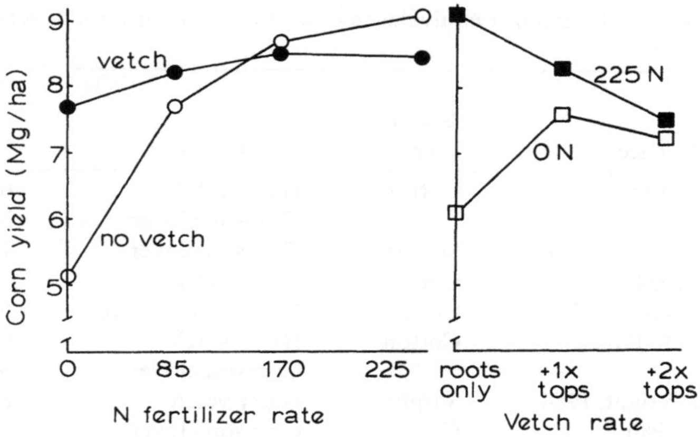

Figure 3. No-tillage corn yield: (A) after vetch cover crop or after winter fallow as a function of N fertilizer rate and, (B) with or without N fertilizer as a function of vetch rate. Vetch rates were obtained by removing top growth from one set of plots (roots only) and transferring this to another set (+2x tops). (S. Corak, University of Kentucky, unpublished data).

been observed that extractable constituents of sweet clover, perhaps coumarin, can inhibit germination and seedling development (McCalla and Duley, 1948). Megie et al. (1967) concluded that ammonia toxicity was the explanation for the poor seedling growth of cotton sometimes observed following alfalfa (*Medicago sativa*). However, relatively few field studies have been able to segregate the effects of specific chemicals from the effects of a generally unfavorable environment (poor aeration, low temperatures, etc.). This is not to imply that allelopathy has not been clearly demonstrated for some specific plants or that it may be of some general importance, only that it has been difficult to investigate in realistic systems.

The effects of cover crops on pests and pathogens, either beneficial or detrimental, could be very important. This is a question that demands further attention.

IV. Nitrogen from Legume Cover Crops

A. Nitrogen Fertilizer Equivalence

As discussed previously, the N contribution from legume winter cover crops is their most commonly observed benefit and almost certainly their primary advantage over grass cover crops. Therefore, determining the magnitude of this contribution is of great importance. *Direct* measurements of N actually transferred from the cover crop to the summer crop are not easily made and limited data are available, not just for legume cover crops, but for any plant N source. This is considered further in the next section.

Table 5. Estimated N fertilizer equivalence of legume cover crops in several recent no-tillage studies

Reference	Summer crop	Cover crop	N fertilizer equivalence (kg ha^{-1})
Brown et al., 1985	Cotton	Hairy vetch	67–101
		Crimson clover	34–67
Eylands and Gallaher, 1984	Sorghum	Crimson clover	75
Varco et al., 1984	Corn	Hairy vetch	78
Mitchell and Teel, 1977	Corn	Hairy vetch + rye	56–112
Touchton et al., 1984	Cotton	Hairy vetch	68
		Crimson clover	68
Breman and Wright, 1984	Sorghum	Hairy vetch	89 or less
Reeves et al., 1986	Corn	Crimson clover	67
Ebelhar et al., 1984	Corn	Hairy vetch	100
		Crimson clover,	50 or less
		Bigflower vetch	50 or less
Worsham, 1986	Corn	Hairy vetch	200
		Crimson clover	100
Buntley, 1986	Corn	Hairy vetch + wheat	56
Flannery, 1981	Corn	Hairy vetch	200
Hargrove, 1986	Sorghum	Hairy vetch	50–128
		Crimson clover	19–128
		Common vetch	30–83
		Subterranean clover	12–103

Many studies provide an indirect measurement of the N contribution by comparing yield response to N fertilizer with and without legume covers. The "nitrogen fertilizer equivalence" of a legume is often calculated as the quantity of fertilizer nitrogen that must be applied to the winter fallow or grass cover treatment to attain a summer crop yield equal to that with the legume cover and no N fertilizer. Several values calculated in this way from recent studies are given in Table 5. These range from approximately 40 to 200 kg N ha^{-1}, but more typically are between 75 and 100 kg N ha^{-1}. Estimating the N contribution in this way is reasonable in a management context, in that the value of the cover crop is assessed in terms of crop yield and fertilizer, easily priced commodities.

However, this approach is misleading for several reasons and it cannot be assumed that the nitrogen fertilizer equivalence value validly estimates any of the following: the quantity of N transferred from legume to summer crop, the difference in N accumulation by summer crops following legumes versus winter fallow, the amount of N fertilizer that will not have to be used if legume covers are, or the amount of N released from the cover crop. A simple experimental problem with this approach is that only three or four widely spaced N rates are generally used, greatly limiting precision of the estimate. Furthermore, many studies report only yield, not N accumulation, which would be a more sensitive indicator of N supply. This approach also implies that all of the yield response is due to the N contribution, while in reality it is clear that the cover crop may have beneficial or detrimental effects even when N is non-limiting, as discussed previously. A more practically meaningful estimate of the N fertilizer replacement value of a legume might be the difference in the amount of fertilizer required to attain optimum yield, or optimum N accumulation, with and without legumes. This is not easily done with most available data sets.

These and other data do suggest that legume residues are a less efficient N source than inorganic fertilizers; that is, the fraction of the total N input that is available to the following crop is smaller for legume N than fertilizer N. This can be seen by comparing the nitrogen fertilizer equivalence values in Table 5 with the values for total legume N accumulated in Table 2. For example, in the Georgia study by Hargrove (1986), the various legumes contained 100–150 kg N ha^{-1} aboveground, but only resulted in as much N accumulation in grain sorghum as 50–100 kg of fertilizer N per ha.

Some earlier work further supports the idea that legume residues are relatively less efficient N fertilizers. Stickler et al. (1959) fertilized corn with several different legumes residues at rates from 71 to 164 kg N ha^{-1} or inorganic N at rates from 0 to 112 kg ha^{-1}. Relative efficiency of legume N was calculated as the quantity of inorganic N required to give the same yield response as the legume, divided by the amount of legume N added. Expressed in this way, legume N was 16–92% as effective as inorganic N, but most treatment fell between 25% and 50%. In a similarly designed experiment, Fribourg and Bartholomew (1956) reached similar conclusions, reporting that the mean efficiency of alfalfa residue N was 34% of that for fertilizer N. However, the ambiguity necessarily introduced by these approaches is illustrated in Figure 4, taken from their data. With low alfalfa N inputs, efficiency was

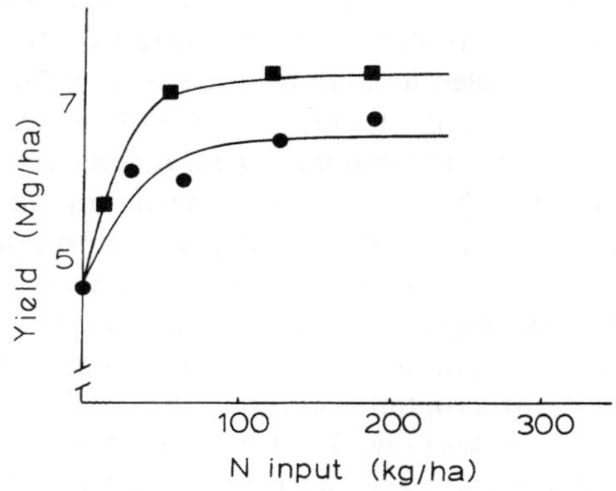

Figure 4. Corn yield as a function of total N input from either N fertilizer (squares) or alfalfa residues (circles). (Adapted from Fribourg and Bartholomew, 1956).

relatively high; at high rates, it was very inefficient. The latter might be due to lower availability of the N in the residues or it might be due to poor plant growth for unknown reasons unrelated to N availability. Visual inspection of their regression lines suggests that maximum crop yields were attained with about the same total N input, regardless of N source (fertilizer or legume). This could be interpreted as indicating that availability of N from the two sources is comparable.

B. Transfer of ^{15}N from Residues to Crops

Some of this uncertainty could be resolved by using ^{15}N-labeled legume residues. Nitrogen isotope techniques have not been widely applied to legume cover crop systems specifically, but there have been several relevant studies of plant uptake of N from labeled residues. This includes one of the first agronomic experiments, of any kind, with ^{15}N (Norman and Werkman, 1943). They added labeled soybean residues to soil in greenhouse pots and determined that 26% of the residue N was accumulated by the plants grown in these soils.

Most studies with labeled residues have used wheat (*Triticum aestivum*) straw or materials of similar composition. Because of the large differences in N content and chemistry between these and legume residues, the results may not be indicative of the fate of N in legume cover crops. Meyers and Paul (1971) incorporated labeled oat residues into microplots in the field. Only 11% of the ^{15}N in these residues was recovered in two successive wheat harvests. Similarly, low transfer was observed in an experiment by Frederickson et al. (1982). Only 9% of the ^{15}N in wheat residues was taken up by the next wheat crop, regardless of whether the residues were plowed in or left on the soil surface. In parallel treatments with labeled fertilizer, recovery was 25–40%. Wagger et al. (1985) reported that 12–33% of the N in wheat residues was recovered in the following sorghum crop.

The low availability of residue N in the studies above might be associated with their high C:N ratio favoring immobilization of N. However, the few experiments done with labeled legume residues also indicate poor recovery. Azam et al. (1985) grew corn in greenhouse pots amended with labeled ammonium sulfate or sesbania (*Sesbania aculeata*). Recovery from ammonium sulfate was only 20%, but N recovery from sesbania was even lower, 5%. Added sesbania reduced uptake of ^{15}N-labeled ammonium sulfate, but the inorganic N had no effect on organic ^{15}N use. These recoveries are unusually low for either fertilizer or organic N, particularly for pot studies.

Yaacob and Blair (1980), in another pot study, measured transfer of ^{15}N from soybean or siratro (*Macroptillium atropurpureum*) residues to Rhodesgrass (*Chloris gayana*). Recovery was low, 13–16%, except with soils that had been cropped to siratro for several years. In this case, up to 56% of the ^{15}N from a single siratro residue addition was taken up by the grass. The explanation for

Table 6. Fate of nitrogen in labeled alfalfa and fertilizer added to soils that were then cropped to wheat

		Percentage of applied N recovered				
	Amount of		In soil after first crop			
Source	N added ($kg\ ha^{-1}$)	In tops first crop	inorganic	organic	Lost after first crop	In tops second crop
Alfalfa	96.7	16.1	5.9	61.3	16.7	4.6
Alfalfa	38.7	18.5	3.7	62.7	15.1	3.8
Urea	50	46.4	3.7	29.2	20.7	3.4

From Ladd and Amato (1986).

this is obscure but may indicate low potential for ^{15}N immobilization in those treatments that were N enriched by cropping history.

In a series of publications, Ladd and his co-workers in Australia have contributed greatly to the understanding of transformations of legume N. In one study, 20–30% of the labeled alfalfa N added to small field plots was utilized by a wheat crop (Ladd et al., 1983). A later experiment compared the fate of alfalfa N with N from various inorganic fertilizer sources (Ladd and Amato, 1986). A portion of these results is summarized in Table 6. Crop uptake of inorganic N was almost three times greater than alfalfa N during the 1st year. In this experiment, alfalfa residues were applied 6 months before planting, but fertilizers were applied at planting. However, since N losses were small under these semiarid conditions, this certainly cannot account for the difference in uptake.

In a Kentucky experiment (Varco, 1986), hairy vetch was labeled by repeated fertilization with ^{15}N-depleted fertilizer. Unlabeled vetch tops were removed from subplots within the experiment described by Ebelhar et al. (1984), where different cover crops had been grown for 7 years. This was replaced with the labeled vetch. With no-tillage management, approximately 20% of the labeled N was recovered in corn grain plus stover. When residues were plowed in this increased to 32%.

C. Decomposition Processes, Mineralization, and Immobilization

The fate and plant availability of N in legume cover crops will be largely determined by the rate and extent of residue decomposition and associated N mineralization. As an example, Huntington et al. (1985) concluded that the mineralization of hairy vetch N in Kentucky was too slow for optimal utilization by no-tillage corn. They suggested that there was "poor synchronization" between N release and potential N uptake and that management practices might need to be adjusted accordingly.

Although there have been few studies of decomposition and N mineralization for legume winter cover crops in conservation tillage systems, there is an extensive literature on plant litter decomposition in general. Thus, we have a relatively good qualitative understanding of the factors regulating litter decomposition kinetics, but little quantitative information applicable to modern cover cropping practices. Without elaboration, we merely list important regulating factors and discuss only some of these as specifically related to some questions of interest. Residue characteristics determining decomposition and mineralization kinetics include N content, C chemistry, particle size, and quantity added. Important soil characteristics include clay content, pH, aeration, and, for low nutrient residues, soil nutrient status. Among environmental or climatic factors, temperature and moisture are surely most significant.

The top growth of legume winter cover crops must be among the most rapidly degraded of plant materials. It is high in protein, low in lignin, generally low in other inhibitory polyphenolic compounds (but see Vallis and Jones, 1973), and has a relatively fine physical structure. Proteins are readily metabolized and assimilated by soil microbes. Stott et al. (1983) measured approximately 50% conversion of protein C to CO_2 in 2 weeks. Actual protein consumption rates were even greater because a large percentage of the protein C was converted to microbial biomass rather than CO_2. This rapid, extensive conversion to microbial biomass was also observed during decomposition of incorporated, labeled *Medicago littoralis* in the field (Amato and Ladd, 1980; Ladd et al., 1981). In these studies, half of the legume ^{14}C was lost in 4–5 weeks and approximately 20% of the remaining ^{14}C and slightly more of the remaining ^{15}N were found in microbial biomass. These results are consistent with the somewhat older ideas of Kuo and Bartholomew (1963). They concluded that plant proteins are rapidly and virtually completely destroyed and that the residual N and C is of microbial origin. If high protein–low lignin residue composition favors retention of C in biomass and microbial products, while high lignin–low protein composition favors retention in humic fractions, as suggested by the results of Stott et al. (1983), will residue chemistry have long-lasting effects on soil organic matter quality or lability? We might speculate, for example, that equal total inputs of C and N as legume residues versus wheat straw plus N fertilizer might lead to equal quantities of soil C and N, but to more labile or active soil C and N with the legume source. This is further discussed below.

An important factor in the application of cover crops to conservation tillage systems is the effect of surface placement compared to incorporation. There is abundant evidence from earlier work that decomposition and mineralization of N are slower for surface residues (McCalla and Russel, 1948; Parker, 1962; Brown and Dickey, 1970). This effect is commonly attributed to lower moisture and nutrient availability for decomposition at the surface. Nutrient limitations for surface residues are only applicable to low-nutrient residues in which soil nutrients are required for decomposition. McCalla and Duley (1943) incubated straw and alfalfa in closed containers, which eliminated the moisture loss problem for surface residues. In this case, incorporated low-N straw decomposed faster than straw on the surface, but placement had no effect on decomposition of the high-N alfalfa. The much slower degradation of residues in the relatively dry environment aboveground, relative to the generally high-moisture environment of incorporated residues is indicated by results of Bartholomew and Norman (1946). They observed much reduced decomposition rates for residues in atmospheres with relative humidity below approximately 90%.

Wilson and Hargrove (1986) measured rates of C and N loss from litter bags containing crimson clover residues. Litter bags were either placed on the surface of no-tillage plots or buried in conventional tillage plots. Approximately half of the N was lost from the buried bags in 2 weeks, yet 8 weeks were required for comparable loss from litter bags on the surface. The C:N ratio remained constant in the surface litter bags, but declined slightly in the buried treatments.

Varco (1986) meticulously collected hairy vetch cover crop residues from no-tillage and plowed field soils at various intervals after the cover was killed (Figure 5). With conventional tillage, the half-life for residue degradation was less than 15 days. With no-tillage, 50% degradation required 45–75 days. The residue half-lives reported in Ladd's Australian studies fall between these Kentucky extremes. In the Kentucky experiments, the more rapid degradation of legume cover crops with conventional tillage was associated with greater legume N uptake by corn, compared to no-tillage (data given above). However, Triplett et al. (1979) concluded that there was no apparent effect of tillage on availability of N to corn planted into alfalfa meadow. Best residue management practices will certainly be climate and soil dependent.

Even though legume residues have a low C:N ratio, it can be expected that a large fraction of their N will be assimilated into microbial cells and retained

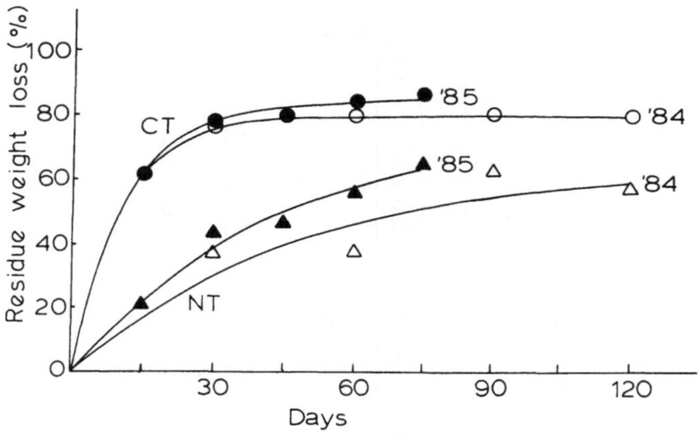

Figure 5. Decomposition of hairy vetch residues in plant-free cylinders in the field. Residues were placed on the surface of undisturbed soil (NT) or were incorporated in cultivated soil (CT). (From Varco, 1986).

Table 7. ^{15}N balance for labeled fertilizer and hairy vetch added to small soil cylinders without plants and incubated in the field

		Percentage of ^{15}N recovered after							
		30 days				120 days			
N source	Tillage[a]	Inorganic	Organic	Residue	Lost	Inorganic	Organic	Residue	Lost
NH_4NO_3	NT	32	8	—	60	12	9	—	79
	CT	35	16	—	50	5	11	—	84
Hairy vetch	NT	8	15	40	37	4	23	21	52
	CT	24	49	13	14	2	33	9	56

From Varco (1986).

[a]NT = no-tillage; CT = conventional tillage.

in the soil. Both Ladd and Amato (1986; Table 6) and Varco (1986; Table 7) measured greater immobilization of legume N than fertilizer N. In both experiments, percent N recovery in organic fractions was at least twice as great for legumes as for fertilizer. In Kentucky, initial immobilization was apparently greater for both N sources in conventional tillage, but between 30 and 120 days apparent immobilized N decreased with conventional tillage yet continued to increase with no-tillage. However, some of the initial organic N in conventional tillage may have simply been undecomposed material that had physically disintegrated to the point where it could not be separated as residues.

D. Residual Nitrogen Availability

This extensive immobilization of legume N and the retention of much of the protein N added to soil in biomass, a presumably labile form, suggest that the long-term residual availability of legume N might be high. Adams et al. (1970, 1973) concluded that the beneficial effect of legumes in rotation on corn nutrition persisted for at least 4 years. Data in Table 8 (Frye et al., 1985) indicate an increasing benefit of hairy vetch crops with time. The ratio of corn yield with vetch to that with winter fallow increased more at low fertilizer N rates, suggesting that this temporal pattern is associated with accretion of available soil N under winter legumes. As discussed elsewhere, relative increases in total soil C and N generally do occur with cover crops. However, at present we know of no good evidence that the residual availability of soil N derived from legumes is any greater than soil N derived from fertilizer, or that high legume N inputs will necessarily increase the fraction of total soil N that is in active pools or labile forms. Fribourg and Bartholomew (1956) observed lower initial availability from alfalfa than NH_4NO_3 (Figure 4). In the 2nd year after application, oats accumulated similar quantities of N from both sources, even though it was probable that more alfalfa N remained in the soil than fertilizer N. Ladd and Amato (1986) also saw about equal N

Table 8. Ratio of no-tillage corn yields with annual hairy vetch cover crops to yields with a winter fallow

Year	Hairy vetch: winter fallow yield with N fertilizer applications of (kg ha^{-1})		
	0	50	100
1977	1.24	1.02	1.24
1978	1.12	0.98	1.13
1979	2.12	1.30	1.22
1980	2.03	1.61	1.43
1981	2.56	1.82	1.69

From Frye et al. (1985).

uptake by the second crop after alfalfa or urea amendment, although twice as much alfalfa N remained after the first crop (Table 6). This limited information suggests that the proportional availability of immobilized N may actually be less for legume than fertilizer sources, but this requires further investigation.

E. Nitrogen losses

Contamination of ground and surface waters by nutrients or organics lost from agricultural soils is a subject of widespread concern. Excessive NO_3^- leaching from cropped land represents not only a potential environmental problem, but a significant waste of a farmer's resources. There are some indications that widespread adoption of conservation tillage might exacerbate this problem. Because less water evaporates in minimum tillage systems, more water is likely to move through the profile, potentially carrying NO_3^- with it (Thomas et al., 1973). On the other hand, there has long been evidence that cover cropping greatly reduces NO_3^- leaching (Jones, 1942; Karraker et al. 1950). Karraker et al. observed that legume covers were as effective as grasses, but this was not consistently the case in Jones's work. Winter cover crops will reduce NO_3^- leaching by depleting both soil NO_3^- and water. This occurs most actively in late winter and spring when the potential for leaching is the greatest in most climates. Varco's data (1986) also indicate that N accumulated in legumes is less subject to leaching than fertilizer during the summer growing season (Table 7). Only the top 20 cm of soil were sampled in this study, which was conducted on a well-drained soil in a fairly wet year, but there was obviously more downward transport of fertilizer than legume N. Losses of N were small in Ladd and Amato's work (1986; Table 6) in a semiarid climate, but slightly more fertilizer N than alfalfa N was lost.

Denitrification losses of legume residue N have not been measured. It might be expected that water removal by a cover crop would minimize denitrification. However, this could be counteracted by the addition of large amounts of readily available substrate to the soil, which would favor gaseous N loss, at least for incorporated residues. The potential for volatilization of NH_3 from legume residues decomposing on the soil surface has not been extensively investigated. Dabney and Bouldin (1985) did measure some NH_3 loss from drying alfalfa hay, so this process may be worthy of further investigation.

V. Effects on Soil Properties

A. Accumulation of Organic N and C

It can be stated as common knowledge that legumes and sod grasses in rotations will increase soil organic matter, or at least maintain it at relatively higher levels than under row crops. Increased organic matter could be beneficial to crop growth by enhancing soil physical properties, water relationships,

Table 9. Soil organic C and N and C:N ratio initially and after 3 years of grain sorghum with various cover crop treatments

Cover crop	Organic C (g/kg)	Organic N (g/kg)	C:N
	0–7.5 cm depth		
Initial	11.3 (a)	0.77(a)	14.7
Winter fallow	7.9 (b)	0.58 (b)	13.6
Rye	8.7 (b)	0.65 (a)	13.3
Crimson clover	8.4 (b)	0.65 (a)	12.9
Hairy vetch	9.7 (c)	0.80 (a)	12.1
	7.5–15 cm depth		
Initial	6.1 (a)	0.49 (a)	12.4
Winter fallow	4.8 (b)	0.37 (b)	13.0
Rye	5.4 (ab)	0.42 (c)	12.8
Crimson clover	4.9 (b)	0.41 (c)	12.0
Hairy vetch	5.5 (ab)	0.51 (a)	10.8

From Hargrove (1986).

Note: Values followed by same letter are not significantly different ($p < 0.05$).

or nutrient reservoirs. The documentation of increased soil organic matter is extensive and only a few relevant examples will be discussed.

Kamprath et al. (1958) measured the effects of oats or hairy vetch winter covers with conventionally tilled corn and various N fertilizer rates on changes in soil C and N over 8 years at four sites in North Carolina. In general, soil organic matter declined without cover crops but tended to increase with either vetch or oats plus N fertilizer. Touchton et al. (1984) concluded that winter legumes with no-tillage cotton caused no measurable changes in soil C or N, but their data indicate strong trends for relative increases with crimson clover or common vetch. Hargrove (1986) measured soil C and N before and after 3 years of no-tillage grain sorghum with several different winter cover treatments in Georgia. A portion of these data is shown in Table 9. Organic matter declined in winter fallow treatments but was generally maintained or declined less with cover crops. The differences were consistent only above 15 cm soil depth.

There is little evidence that soil organic matter accumulation is highly sensitive to type of cover crop or residue used. Legumes do not consistently result in more or less soil organic C than equivalent quantities of higher C:N materials, such as grass or wheat straw. Larson et al. (1972) added different rates of different residues to soil for 11 consecutive years. For a given mass of residue, soil C accumulation was comparable for legumes, straw, and even sawdust. Soil N increases were also surprisingly similar for all materials except sawdust—an extremely low-N substrate. Kamprath et al. (1958) observed no consistent differences in soil C and N between hairy vetch and oats if adequate fertilizer N was supplied for good crop growth. Hargrove (1986; Table 9) noted that rye covers resulted in just as much soil N accumulation as crimson clover, and at least as much soil C, even though the residues of the former contained less than one fourth as much N as the latter. This indicates that the retention of organic N in soils is dependent on the inputs and retention of organic C. However, hairy vetch, which contained approximately as much N and somewhat less C than crimson clover, resulted in significantly greater soil C and N. Both legume covers caused a reduction in soil C:N, hairy vetch more so than crimson clover. The interactive effects of substrate quality, soil properties, and quantities of C and N inputs on soil organic matter are largely unexplained. Well-designed, long-term experiments with precise accounting of C and N inputs are needed.

The effect of tillage, or no-tillage, on long-term retention of cover crop C and N in soil is also uncertain. Although it is quite clear that soil organic matter will be relatively increased as tillage is reduced (e.g., Blevins et al., 1983) and that short-term residue decomposition loss is slower with no-tillage (as discussed elsewhere), it is less certain that cover crops will cause a greater difference in soil organic matter, versus winter fallow, in no-tillage than in plowed systems. Albrecht (1936) either turned in or left clover residues on the surface of field plots for 15 years. The method of application had little effect on final soil N, with only 3% more of the applied N being retained in the mulched treatment. Beale et al. (1955) observed more soil N after 10 years with minimum tillage cover crops than moldboard-plowed cover crops. However, the only winter fallow treatment included was plowed. Such observations may reflect greater loss of soil organic matter with greater tillage, and not relatively less effect of cover crop inputs on organic matter in plowed systems than minimum tillage systems. Data from Utomo (1986) in Kentucky do show a greater difference between organic C in hairy vetch and winter fallow treatments for no-tillage than for conventional tillage (Table 10). Vetch had a small effect in conventional tillage, but a significant effect in no-tillage. This

Table 10. Total soil C from 0 to 7.5 cm as affected by N fertilizer rate, tillage, and cover crop

	N rate (kg ha^{-1})			
	0		170	
Cover crop	CT[a] (g kg^{-1})	NT[a] (g kg^{-1})	CT[a] (g kg^{-1})	NT[a] (g kg^{-1})
Winter fallow	15.7 d	19.3 c	16.8 d	23.7 b
Hairy vetch	16.0 d	22.0 b	16.9 d	25.2 a
Rye	16.3 d	19.8 c	15.2 d	22.4 b

From Utomo (1986).

Note: Values followed by same letter are not significantly different ($p < 0.05$).

[a]CT = plowed and disked; NT = no-tillage.

experiment had been entirely in no-tillage for several years then half of each plot was spring plowed for 2 years before these measurements were made. As expected, all no-tillage treatments contained more total C and N than the analogous plowed treatment. It should be noted that the tillage by cover crop interaction was observed less consistently on earlier sampling dates and in measurements of organic N.

B. Soil Water

Legume winter cover crops can influence soil water relationships through at least four distinct mechanisms: (1) reduced evaporation due to the mulch effect, (2) increased transpiration from the cover crop, (3) increased infiltration and retention of precipitation, and (4) altered water use by the summer crop if its growth is affected. Most investigations have dealt with only one of these at a time, most commonly the mulch effect. The last mechanism has not been investigated at all, to our knowledge. The third is indirectly considered here within the discussion of soil physical properties. The synthesis of these interacting and potentially counteracting mechanisms to predict the net effect of cover crops on water availability to summer crops represents a formidable challenge. There is an apparent opportunity for the application of simulation modeling to this problem, but this approach has not yet been exploited.

The nature of the mulch effect in retarding water evaporation has been discussed in detail by Phillips (1984) and Bond and Willis (1969) among others. Briefly, the lower rates of water loss under a mulch are attributed to: (1) the barrier to water vapor movement provided by the mulch that has the effect of reducing the rate of conversion of liquid water to vapor at the soil surface; (2) essentially shading, whereby direct solar radiation reaching the soil surface is reduced; and (3) insulation, whereby conduction of heat from the air to the soil surface is reduced. It is clear that mulch reduces the initial rate, but not the final extent of soil water loss. The greatest differences in water contents between mulched and bare soils can be expected during short dry periods (7–14 days according to Bond and Willis [1969]), not long ones. Mulch effects will be most important early in the growing season because virtually complete shading is provided by a closed crop canopy.

The higher moisture contents repeatedly observed in comparisons of no-tillage with plowed soils have been attributed primarily to the mulch effect (Blevins et al., 1971; Lal, 1974; Nelson et al., 1977; Gantzer and Blake, 1978). With regard to water conservation (perhaps in contrast to erosion control, as discussed later), lots of mulch seems to be significantly better than a little. Thus, the additional early season residue cover provided by a winter cover offers important advantages even in minimum tillage systems where there is a considerable accumulation of surface residues from the summer crop. Unger (1978), in Texas, reported increases in soil-available water at sorghum planting with mulch rates up to 12 Mt/ha. The increase was 0.68 cm of water per Mg of mulch per ha. In Ohio, Triplett et al. (1968) measured total available water

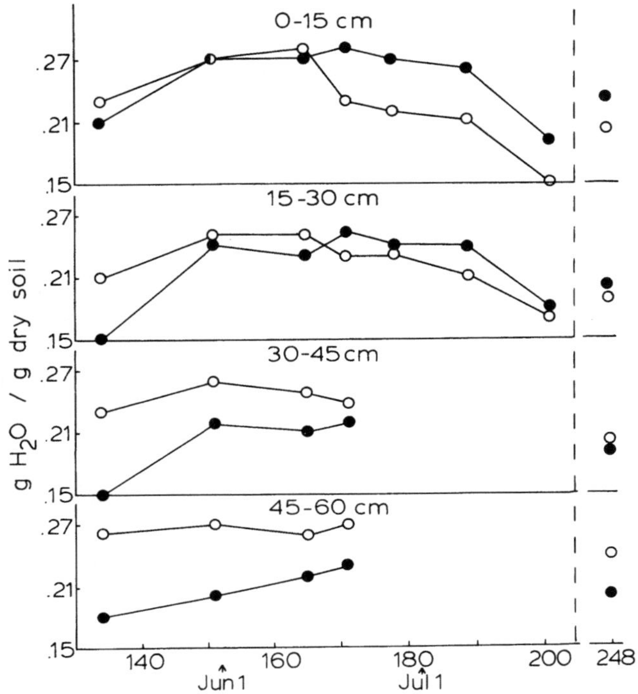

Figure 6. Gravimetric soil water content by date and soil depth for no-tillage soil planted to corn following vetch (filled symbols) or winter fallow (open symbols). (After Utomo, 1986.)

from mid-June to mid-September of 43, 44, 53, and 59 mm for plowed soil, no-tillage residue removed, no-tillage normal residue, and no-tillage double normal residue treatments, respectively.

The additional conservation of water due to a mulch from hairy vetch winter cover relative to corn stalk residues alone in no-tillage corn is shown also in Figure 6 (Utomo, 1986). At the soil surface in June and July, gravimetric water contents were higher with hairy vetch. These Kentucky data are of further interest because they also document the importance of the second mechanism discussed above, increased transpiration of soil water by the cover crop before it is killed in the spring. Significant depletion of soil water in the hairy vetch treatment was observed at depths to 60 cm, the deepest measured, at corn planting time in May. Near the surface, this depletion was quickly counteracted by recharge and reduced evaporation. However, at greater depths, the lower water contents under vetch persisted well into the growing season. Rye cover crops were generally no different from vetch in this study.

The net result of reduced evaporation versus increased transpiration, and its impact on growth of the summer crop, obviously will be weather and site specific. While we have occasionally observed poor early season corn growth attributable to water depletion by the cover crop, in cold or poorly drained soils this accelerated spring drying would be advantageous.

C. Soil Temperature

A surface mulch will generally cause lower soil temperatures, because the mulch reduces energy transfer to the soil. A study by Van Wijk et al. (1959) elegantly demonstrates that this will be beneficial to crop growth in some locations and detrimental in others. They measured soil temperatures and early season corn growth in mulched and bare soil. The same experiment was done in Iowa, Minnesota, Ohio, and South Carolina. Mulch reduced early season soil temperature, particularly daily maxima, in every case, but this reduced corn growth only in the northern locations. In the tropics, temperature reductions due to mulching are perceived as being desirable (Lal, 1976; Lal et al., 1980). Ghuman and Lal (1982) in West Africa reported that rice (*Oryza sativa*) straw mulch reduced daily maximum temperatures by as much as 16° C. Wade and Sanchez (1983) measured lower temperatures and better moisture relationships in mulched than unmulched soil in the Amazon Basin. Without fertilizer, crop yields were increased by mulch, but not with fertilizer. This suggests that the mulch benefits are due to nutrient supply, not necessarily to more favorable temperature.

In Kentucky, Knavel and Herron (1981) associated their lower cabbage yields with no-tillage, versus plowed, with lower soil temperatures. Hoyt and Hargrove (1986) also noted lower temperatures in strip-tilled than in cultivated plots. Cover crops caused small further temperature reductions over bare plots with the same strip-till management. A transient inhibition of potato emergence was apparently associated with these lower temperatures.

Utomo (1986), in Kentucky compared soil temperatures in plowed and no-tillage corn plots that had rye, hairy vetch, or no winter cover crop. Tillage was the only variable that had a statistically significant effect, with plowed soil being warmer. As shown in Table 11, there was a trend for daily maximum

Table 11. Soil temperatures at 5 cm depth between corn rows for no-tillage with hairy vetch cover crops and with no cover crops, corn stalk residue only

Treatment[a]		Mean daily maximum or minimum (°C) for time periods				
		5/23–5/27	5/23–6/3	6/4–6/10	6/11–6/17	6/18–6/24
CS	Min	16.2	13.1	17.7	19.8	21.0
	Max	23.7	21.3	27.7	30.8	30.3
HV	Min	16.0	13.4	18.5	19.8	20.5
	Max	21.5	20.6	26.5	29.3	29.4

From Utomo (1986).

[a] CS = corn stalk residue only; HV = hairy vetch cover crop.

temperatures with hairy vetch no-tillage to be lower than for winter fallow no-tillage. Yet these differences probably are not large enough to have a measurable impact on crop growth or soil processes. Mitchell and Teel (1977), in Delaware, report 1–7°C lower temperatures under cover crop mulches than under bare soil at 2.5–10.2 cm soil depth. Temperatures in the mulch were actually higher than at the bare soil surface, indicating restricted energy transfer across the mulch-soil interface.

A change in albedo due to a cover crop might occasionally be large enough to influence significantly soil temperature (McCalla, 1959). Darker colored mulches, such as hairy vetch, would be expected to absorb somewhat more energy from solar radiation than lighter grass residues or some soil surfaces. One additional physical effect that should be considered is that transpiration by the cover crop before summer planting will reduce the effective heat capacity of the soil. This could result in warmer soil temperatures in the spring with a cover crop, although we are aware of no data documenting this possibility.

D. Soil Physical Properties

It is reasonably well documented that sod crops will promote more favorable physical conditions than row crops or fallow conditions. For example, Hovis (1943) reported that aggregation, as determined by wet sieving, was greater in orchards with sod ground covers. Strickling (1950) also observed that larger aggregate sizes were favored in hay plots than with corn or soybeans. Jordan et al. (1956) found decreased bulk density, greater porosity, aggregation, and intrinsic permeability in soils that had been in kudzu (*Pueraria labata*) for 4 years relative to continuous corn. Root development of corn following kudzu was enhanced in this soil with a genetic pan. Uhland (1949) also emphasized the value of deep-rooted legumes such as kudzu and alfalfa. In Africa, 2–3 years of grass or legume cover increased infiltration rate and porosity on an eroded soil with initially poor physical properties (Wilson et al., 1982).

Including sod grasses or legumes as winter cover crops or in short-term rotations with summer grains can apparently have similar benefits to long-term sods. However, much of the information on this is derived from studies with conventional tillage where residues are incorporated. Plowed down winter covers resulted in greater plasticity and lower bulk density than winter fallow in Virginia experiments by Obenshain and Gish (1941). Legumes, including cowpeas, vetch, crimson, and red clover, were more effective than grass covers in this study. In contrast, grass (redtop) was more beneficial than red clover in promoting aggregation in a potato rotation (Wisniewski et al., 1958; Saloman, 1962). The degree of aggregation declined over 8 years in plowed corn fields with no cover crop, but stayed constant when a cover crop of rye and hairy vetch was grown every winter and plowed under (Beale et al., 1955). Benoit et al. (1962) reported that a rye winter cover crop enhanced aggregate stability and hydraulic conductivity of surface soil, but this effect was only detectable after 3 years of the rotation. Penetrometer resistance was lower when a hairy vetch cover was included in cotton/corn and cotton/peanut (*Arachis hypogaea*) rotations and this was believed to be related to yield increases (Welch et al., 1950). Different results were obtained by Siddoway (1963); aggregation was actually lower in wheat with alfalfa or sweetclover than in wheat/fallow rotations. Also, Elson (1941) noted little difference in aggregation among wheat-clover, wheat-corn, and corn-hay rotations, although clover/hay did measurably enhance aggregation. In assessing the literature on soil physical properties, it should be pointed out that many investigators have questioned the sensitivity or the relevance of the commonly used techniques (e.g., Allison, 1968) and that field variability often can make it impossible to determine treatment effects.

Little recent soil physical data are available on legume winter cover crops in no-tillage or conservation tillage systems. Brown et al. (1985) observed no significant difference in bulk density and water infiltration rates for hairy vetch or crimson clover covers versus winter fallow with no-till cotton. In another experiment, Touchton et al. (1984) did record faster water infiltration on plots with these legumes compared to fallow (Figure 7). In Kentucky, the authors measured steady state ponded infiltration rates of plots that had hairy vetch or no cover for 7 years. The former was 88 and the latter 71 mm hr^{-1}, but spatial variability was so great that this difference was not significant (unpublished data).

Do cover crops in no-tillage have the same effects on soil structure as incorporated cover crops? Will harvesting cover crop top growth eliminate or reduce the enhancement of soil structure? Is it the root growth or the added organic matter input of cover crops that enhances structure? Answering these questions is hindered by our minimal understanding of the basic mechanisms of soil aggregation. Several years ago, Allison (1968) concluded that this is one of the least understood subjects in soil science, and it is still difficult to disagree with this conclusion. Allison noted that aggregation is a two-step

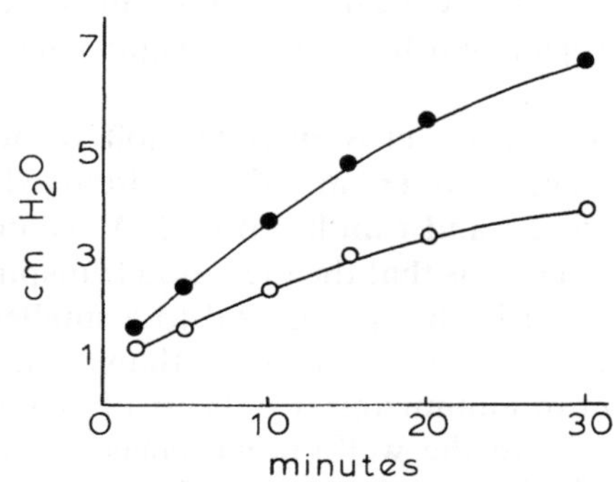

Figure 7. Water infiltration on soil with hairy vetch cover crops (filled symbols) or with no cover (open symbols). (Adapted from Touchton et al., 1984.)

process, involving a largely physical step of putting particles together, followed by a chemical cementing of these particles by organic or inorganic agents. He proposes that plant residues as well as humic compounds, per se, have little or no role in aggregation. Rather, it is the microbial decomposition products of these, particularly polysaccharides and perhaps microbial cells themselves, which act as the cementing agents. Others have made similar conclusions (McCalla, 1945; Chepil, 1955). From this, Allison concluded that organic matter inputs are relatively unimportant to aggregation and that their effects are rather short-term. He atrributes the benefits of sod crops primarily to root-associated aggregation. Some field results also emphasize the important effect of rooting on soil structure (Uhland, 1949, Wisniewski et al., 1958). Yet it is clear that organic matter additions alone can enhance aggregation (Browning, 1937; Van Doren and Stauffer, 1943, McCalla, 1945). Hovis (1943) concluded that surface mulches were actually more effective than sods at promoting aggregation. Inert surface mulch did not enhance soil structure, so the effect of plant residues must be associated with their ability to support microbial growth, not with the physical protection or insulation of the soil surface. It might be expected that surface residues would lead to microbial growth and polysaccharide production above the soil rather than in it. However, Siddoway (1963) indicates that surface residues promoted aggregation more than incorporated residues. Obviously, there is much yet to be explained about soil structure and how it is affected by management practices.

VI. Erosion Control

Legumes and winter cover crops have long been recommended as an effective method of reducing soil erosion (Pieters et al., 1950; Hendrickson et al., 1963; Brill et al., 1963). Considering the current level of concern about soil degradation by erosion and also about adverse effects on water quality through losses of sediment and agricultural chemicals in runoff, this potential advantage of legume winter cover crops might be ranked in significance as high or higher than any other advantages.

Legume cover crops can reduce erosion and runoff by three possible mechanisms. First, if soil structure is enhanced (as discussed elsewhere), water infiltration can be increased and runoff reduced. A second mechanism that has less often been considered, is that the increased transpiration from a cover crop will deplete soil water in the spring, and to a smaller extent in the fall, and thereby favor water movement into, rather than over, the soil. This could be extremely important in minimizing the effects of heavy early season precipitation. Finally, there are the well-known effects of a surface mulch or standing vegetation in dissipating the energy of raindrops, protecting surface soil, and reducing the velocity of that water that does move over the surface. It is not possible to evaluate the relative contribution of these mechanisms, but reports that cover crops reduce erosion even when they are plowed down (Beale et al., 1955) suggest that surface effects are not the sole factor.

When legume cover crops are employed in minimum tillage systems, it is expected that the greatest erosion control benefit might be derived from the conservation tillage practices rather than the cover crop itself. This may be particularly true in no-tillage where all of the residues of the summer crop remain on the surface. When corn or sorghum is grown and winter temperatures are not high, the grain crop residues alone may provide an adequate mulch throughout the year. The two studies illustrated in Figure 8 indicate that less than 1 Mg of mulch per ha greatly reduces soil loss and that rapidly diminishing returns are observed at higher mulch rates. These measurements were made with added straw mulch and so are not directly comparable to many field practices. However this does suggest that the annual production of 4–11 Mg of corn residues per ha (Kitur, 1982) will provide an effective mulch.

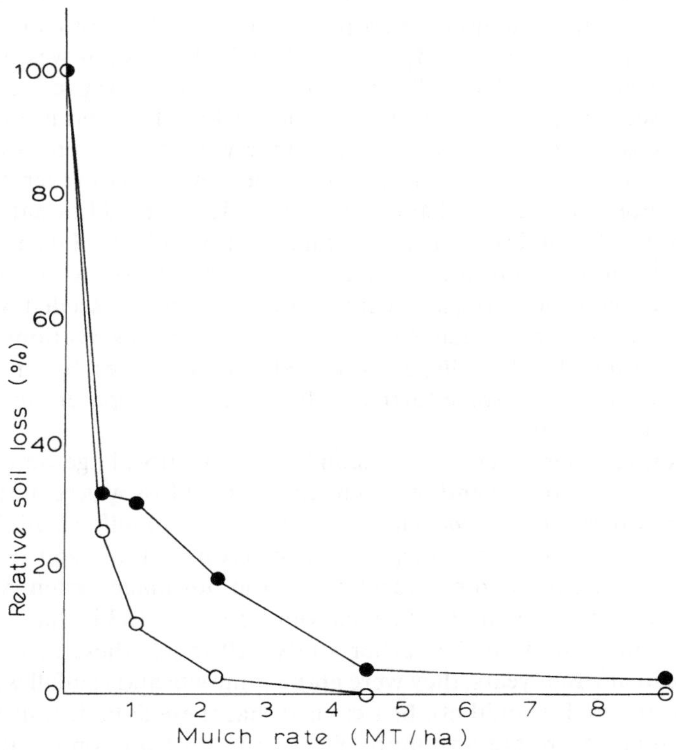

Figure 8. Relationship between quantity of mulch and soil loss in rainfall simulator studies. Soil loss is expressed as percentage of loss with no mulch, which was 27.8 mg ha^{-1} for Mannering and Meyer (1963) (open symbols) and 62.3 mg ha^{-1} for Meyer et al. (1970) (filled symbols).

Table 12. Average annual runoff, soil loss, and herbicide loss in cornfields with various treatments

Treatment[a]	Runoff cm	Soil loss Mg ha^{-1}	Cyanazine loss % of applied
CT	5.71	14.22	2.64
NT-CS	0.54	0.36	0.27
NT-V	0.21	0.02	0.10

After Hale et al. (1984).

[a] CT = Conventional tillage; NT-CS = no-tillage with only a corn stalk mulch; NT-V = no-tillage with interseeded crown vetch.

This is supported by Frye et al. (1985), who used the universal soil loss equation to predict the extent of erosion control from no-tillage and from winter cover crops. The calculated losses for conventional tillage were 18 Mg ha^{-1} yr^{-1}, twice the tolerance limit. For no-tillage without a cover crop, soil loss was 2.2 Mg. If a winter cover crop was added to the no-tillage system, a slight further reduction to 2.0 Mg was predicted. However, it was noted that the use of cover crops with no-tillage would allow corn to be produced on this soil with slopes of up to 20%, keeping erosion below the tolerance limit.

Direct measurement of the effects of legume winter cover crops on runoff and erosion in minimum tillage have not yet been made. However, two series of investigations do indicate that this will provide some additional benefit in erosion control beyond the benefits of conservation tillage. Hale et al. (1984) measured erosion, runoff, and cyanazine loss in plowed corn fields, with no-tillage and no cover crop, and with no-tillage corn interseeded with a live perennial mulch of crownvetch (*Coronilla varia*). Their observations are summarized in Table 12. No-tillage alone reduced all losses by an order of magnitude or more, but some further reductions were apparent in the interseeded vetch treatment.

In a Georgia watershed study, erosion losses from no-tillage were less than 1% of conventional tillage and over-winter sediment loss was reduced 25-fold with a winter barley (*Hordeum vulgare*) crop relative to fallow soil (Langdale et al., 1979). More relevant to legume cover crops, annual soil losses from winter crimson clover/summer sorghum in the no-tillage system were only 4.3 kg ha^{-1} (Hargrove et al., 1984). This compares with 133 kg ha^{-1} for no-till sorghum/winter barley in the earlier study. Although these measurements were made in different years, they were at the same site and rainfall was similar in the two studies. It should not be assumed that these data demonstrate that legume winter covers are any more effective in erosion control than grass winter cover crops if both are planted and harvested in the same manner and at the same time.

The additional surface protection offered by a legume mulch may be small in many conservation tillage fields, yet this may become quite important where residues decompose rapidly, where residue production is low (as in soybeans or cotton), or where steep slopes or other factors make the erosion hazard unusually high.

VII. Economics

While legume winter cover crops may be attractive to agronomists for a variety of reasons (soil protection and improvement, environmental quality benefits, decreased reliance on synthetic N fertilizers and fossil fuel), their appeal to and acceptance by producers will be determined primarily by economics. Will they increase profits? No definite or general answer is possible since responses to legume cover crops are variable. Economic analyses are further complicated by the impossibility of precisely assigning a cost for soil erosion or environmental degradation or a value for long-term increases in soil productivity.

If legume cover crops are considered to be only a replacement for N fertilizer, then the analysis is greatly simplified and reduces to a question of the relative cost of seeding the legume versus the cost of buying and applying an equivalent quantity of fertilizer. This situation is exemplified in a study by Flannery (1981). With hairy vetch cover, yield of no-tillage corn for silage was independent of N rate and approximately equal to the yield with 202 kg of N fertilizer per ha and a rye cover crop. A comparison of the cost of hairy vetch with this quantity of fertilizer shows a slight advantage to the vetch without fertilizer treatment, approximately a 10% higher profit than with rye and 202 kg of N. However, the N fertilizer equivalence of the vetch in this study was higher than in most cases, as discussed previously.

Frye et al. (1985) report the cost of hairy vetch seed to be \$1.54 kg^{-1}. At a recommended seeding rate of 40 kg ha^{-1}, this amounts to \$61.60 ha^{-1}. They further report the cost of N fertilizer to be \$0.60 kg^{-1}. If vetch replaces 100 kg N ha^{-1}, a typical value, then the \$60.00 cost of fertilizer is roughly equal to the cost of vetch seed. This ignores the cost of applying seed or fertilizer and aerial seeding would generally be more expensive than fertilizer application. But in general it can be concluded that the current costs of establishing legume cover crops are approximately equal to the cost of the N fertilizer that they can be expected to replace.

It is interesting that Coleman (1941) reached the same conclusion more than 40 years ago. He estimated the cost of the legume cover at \$8.64 ha^{-1}, but his data suggested that this could replace only about 30 kg N ha^{-1}. (This lower value might be attributed to the rather poor yields of both the cover crop and the following summer crop in this study.) It is evident from these numbers that over the last 5 decades the cost of legume seeding has increased, as has the cost of N fertilizer. Although recent memories might suggest that N fertilizer prices of the future will increase significantly, this cannot be assumed to favor the profitability of legume cover crops unless seeding costs increase

Table 13. Returns above operating costs for no-tillage corn with hairy vetch or with no cover crop at two locations in Kentucky

		N rate (kg/ha^{-1})		
Location[a]	Winter cover	0	50	100
		net return $/ha		
Lexington	Hairy vetch	148	146	269
	Corn stalks only	7	74	158
Princeton	Hairy vetch	−96	−57	35
	Corn stalks only	−195	−121	32

From Frye (1986a).

[a] Means of 5 years at Lexington, 6 years at Princeton.

less. The relative profitability of legume covers would be greatly increased if seeding rates could be reduced without significant sacrifices in production. It is probable that this may be the case under at least some circumstances. The relatively high rates recommended were generally developed for forage production systems and little recent work has evaluated seeding rate in modern cover crop systems.

Detailed economic analyses for legume cover crop experiments in Kentucky have been published (Frye et al., 1985; Frye, 1986a) and some of the results are summarized in Table 13. These results differ from those of Flannery (1981) in that hairy vetch treatment yields exceeded those without a cover crop, even at the highest N rate. This might indicate that the cover crop is providing benefits other than N supply or it may be that at higher N fertilizer rates the yields of the two treatments would have converged. The conclusion in this case, most evident at the Lexington site, is that N fertilizer rates should not be reduced when hairy vetch is grown. The most profitable treatment is hairy vetch, not because it reduces fertilizer costs but because it increases yields and gross returns. Profits at the Princeton site were negative or low because of 2 drought years in 6 and, of course, depressed corn prices. Nevertheless, the hairy vetch treatment was still superior.

A study of Touchton et al. (1982) offers two strategies for improving the profitability of legume cover crops. They allowed the legumes to grow long enough to form seed, saving the cost of reseeding in subsequent years. Perennial winter legumes and self-reseeding are discussed briefly in the next section. Another alternative is to harvest the legume top growth. Winter legumes will generally make high-quality feed for hay or grazing. Touchton et al. (1982) found no difference in the yields of sorghum when top growth was removed or left. However, there were relatively small responses to applied N fertilizer for all treatments and the N contribution from the legumes may not have been an overriding yield-controlling factor in this study. The limited data available suggest that less than one fourth of the N accumulated by legumes is below

Table 14. Effect of cover crop removal and N rate on yield of no-tillage corn grain

N rate ($kg \cdot ha^{-1}$)	Cover crop management	Corn yield with Crimson clover ($kg \cdot ha^{-1}$)	Corn yield with Hairy vetch ($kg \cdot ha^{-1}$)
0	removed	5,710	4,520
0	mulch left	8,150	6,330
180	removed	6,900	6,710
180	mulch left	8,340	7,780

From Wilkinson and Dobson (1981).

ground (Jensen and Frith, 1944; Mitchell and Teel, 1977). However, not all of the aboveground N would be removed by cutting or grazing. Martin and Touchton (1982) found that the aboveground N content of four winter legumes averaged 45, 40, and 100 kg ha^{-1} for grazed, hayed, and undisturbed treatments, respectively. To obtain the same grain sorghum yields as on the undisturbed plots with no N fertilizer, 34 kg N ha^{-1} were required on the cut or grazed plots. Wilkinson and Dobson (1981) in Georgia also showed significant reductions in N availability to corn when hairy vetch or crimson clover covers were cut and removed before planting (Table 14). Although harvesting the legume will generally mean sacrificing some of the N benefit, in some cases the legume protein will be more valuable as a feed than as a fertilizer. Also, some of the other benefits, erosion control and possible enhancement of soil structure, will be partially maintained. The availability of the option to use the legume in various ways would be advantageous for many farmers.

VIII. Perennial Legume Covers

Perennial or self-reseeding legumes offer an attractive solution to the cost and labor required to establish winter cover crops annually. They also present some very difficult management problems, primarily due to competition between the summer species and the cover species. That this competition is mainly for water is indicated by a study of Box et al. (1980) with winter grass covers and corn in Georgia. When only strips of the mulch were herbicide-killed, nonirrigated corn yields were dramatically reduced relative to completely killed mulches. For irrigated corn, yields with killed and partially living mulches were comparable.

Only a few investigators have attempted to develop workable techniques for perennial legume covers, with variable success. As mentioned above, Touchton et al. (1982) allowed crimson clover to grow until June in Georgia, by which time it has formed seed. Grain sorghum was then no-till planted into the mulch. Production of sorghum and regrowth of crimson clover in subsequent years were both good. The limitation to this approach is that for other summer crops and in other regions, delaying the planting date until the legume forms seed would cause significant yield reductions.

Elkins et al. (1983) reported success with living grass mulches suppressed by herbicide application and planted to corn or soybeans. They were able to maintain the perennial mulch, yet minimize competition during the summer. With alfalfa and other legumes, they were unable to suppress adequately the mulch and at the same time keep it capable of regenerating in the fall.

Vrabel et al. (1981) reported increased yields of sweet corn planted into living mulches of white, ladino, or red clover or alfalfa relative to no cover crop. Only 19 kg N ha^{-1} were applied in this experiment. To eliminate competition, they used a combination of mowing, multiple herbicide applications, and rototilling of a 0.34-m strip into which the corn was planted.

Hartwig (1974) did not rely on unusually heavy or frequent herbicide applications or other expensive methods of suppressing the cover crop. But in this case, the corn was irrigated to minimize competition for water. With a living crownvetch mulch and 105 kg N ha^{-1}, he obtained corn yields of 10.35 Mg ha^{-1}.

In Kentucky, the authors have grown nonirrigated corn in established stands of bigflower vetch and crownvetch (Juaregui, 1986). An advantage of bigflower vetch is that it completes its life cycle and reseeds earlier than most legumes. Once mowed and/or killed, it should offer little competition to the summer crop and there should be little or no regrowth until seed germinates during cool autumn weather. A significant disadvantage of bigflower vetch is that it generally does not grow as well or accumulate as much N as hairy vetch (Ebelhar et al., 1984). Crownvetch is another species that should be relatively easily suppressed during the warm growing season, but will behave as a true perennial rather than reseed. Corn could be planted in mid-May in these systems. In this experiment, a combination of mowing and multiple herbicide applications controlled the legume and minimized competition. This was indicated by similar soil water contents throughout the season in fallow and cover crop plots. Corn yields, shown in Table 15, were similar for the cover crop and fallow treatments. However, there was no evidence of a N contribu-

Table 15. Yield of no-tillage corn grain with different N fertilizer rates and crownvetch, bigflower vetch or no cover

	Corn yield with		
N Rate (kg ha^{-1})	Crownvetch (Mg ha^{-1})	Bigflower vetch (Mg ha^{-1})	No cover (Mg ha^{-1})
0	4.3	4.5	4.9
85	6.0	6.6	6.3
170	7.4	7.0	7.4

From Juaregui (1986).

tion from the legumes; yields and N content of corn were similar at all fertilizer rates. We have no ready explanation for this.

In addition to the precise management and often the extra inputs required to control competition, initial establishment of perennial legumes is usually slow and difficult. Several months or even a full year may be needed to establish sufficiently vigorous stands. This could mean sacrificing an income-producing crop. These problems make it unlikely that perennial legume covers will be widely used in the near future. Yet their potential advantages are great enough to justify further research, and it is hoped that this will lead to practical variety-herbicide-management systems in the future.

IX. Conclusions

The most studied benefit of legume winter cover crops is their contribution of N to the summer crops that follow. It is clear that in a favorable climate, with proper management, winter legumes can accumulate well over 100 kg N ha^{-1} without significantly interfering with the scheduling of a summer grain crop. It appears that most of this N will be derived from fixation. It is also well established that this will contribute significantly to the N nutrition of the following crop. Indirect estimates generally indicate that cover crops can supply as much N as 75–100 kg of synthetic fertilizer N per ha. Many scientifically interesting and important questions regarding the fate of legume residue N, and organic N inputs in general, have not been adequately addressed. It can be expected that future research, particularly with isotopically labeled residues, will provide more insight about the rate of release and uptake, and the extent of immobilization and losses from these sources. Additional experiments comparing the fates of N from legume versus inorganic fertilizer will be of considerable value.

Even less is known about effects of legume winter covers other than those related to N. Effects on water availability actually may be more significant than N in determining summer crop yield, since the N contribution of a legume can be more easily and economically replaced than a water contribution. However, since winter legumes can both deplete and conserve soil water, by early season transpiration and by the mulch effect, respectively, predicting the net effect on water relations of the summer crop will not be a simple matter. Such effects will surely be climate and site specific.

It might be expected that including legume winter cover crops in continuous row crop systems would better maintain or increase soil productivity over the long term. Accumulation of soil organic matter and enhancement of physical properties could eventually provide more agronomic benefit than the first year N contribution, and so offer an even more convincing justification for the adoption of legume cover cropping. Yet the long-term experiments necessary to document such effects have not been conducted.

The environmental consequences of winter cover crops could also be of great importance, but these also have not been thoroughly investigated. It is evident that legume covers will reduce soil erosion and minimize runoff during spring and winter. But, quantitatively, how important is this additional protection in minimum tillage systems? Winter cover crops seem to offer a tremendous opportunity to reduce leaching from cropped lands. This is currently a subject of widespread discussion and national concern, yet we have little recent data on the effectiveness of this practice.

It is interesting to note the marked similarities between two lists of needed research on legume cover crops compiled by Hargrove in 1981 and by Pieters and McKee in 1929. Both lists consist primarily of: (1) selection of adapted legumes, (2) development of cover crop seeding and management practices, (3) characterizing effects on soil water relationships, (4) understanding effects on soil physical and chemical properties, (5) measuring effects on erosion and leaching, (6) quantifying N availability and cycling with legume N sources, and (7) investigating effects of winter legumes on pests and diseases of summer crops. This might be taken as an indication that little was learned in the intervening 52 years. But it also suggests that legume winter cover crops were relatively ignored during this era. During the last few years, it has become apparent that there is an opportunity to use this very old practice in modern cropping systems.

References

Adams, W.E., H.D. Morris, and R.N. Dawson. 1970. Effect of cropping systems and nitrogen levels on corn (Zea mays) yields in the Southern Piedmont Region. *Agron. J.* 62:655–659.

Adams, W.E., H.D. Morris, J. Giddens, R.N. Dawson, and G.W. Langdale. 1973. Tillage and fertilization of corn grown on lespedeza sod. *Agron. J.* 65:653–655.

Albrecht, W.A. 1936. Methods of incorporating organic matter with the soil in relation to nitrogen accumulations. *Research Bulletin 249*, University of Missouri Agricultural Experiment Station.

Allison, F.E. 1968. Soil aggregation—some facts and the fallacies as seen by a microbiologist. *Soil Sci.* 106:136–143.

Amato, M., and J.N. Ladd. 1980. Studies of nitrogen immobilization and mineralization in calcareous soils: V. Formation and distribution of isotope labelled biomass during decomposition of ^{14}C and ^{15}N-labelled plant material. *Soil Biol. Biochem.* 12:405–411.

Azam, F., K.A. Malik, and M.I. Sajjad. 1985. Transformations in soil and availability to plants of ^{15}N applied as inorganic fertilizer and legume residues. *Plant Soil* 86:3–13.

Baldock, J.O., R.L. Higgs, W.H. Paulsen, J.A. Jackobs, and W.D. Shrader. 1981. Legume and mineral N effects on crop yields in several crop sequences in the Upper Mississippi Valley. *Agron. J.* 73:885–890.

Bartholomew, W.V., and A.G. Norman. 1946. The threshold moisture content for active decomposition of some mature plant materials. *Soil Sci. Soc. Am. Proc.* 11:270–279.

Beale, O.W., G.B. Nutt, and T.C. Peele. 1955. The effects of mulch tillage on runoff, erosion, soil properties and crop yields. *Soil Sci. Soc. Am. Proc.* 19:244–247.

Benoit, R.E., N.A. Willits, and W.J. Hanna. 1962. Effect of rye winter cover crop on soil structure. *Agron. J.* 54:419–420.

Blevins, R.L., D. Cook, S.H. Phillips, and R.E. Phillips. 1971. Influence of no-tillage on soil moisture. *Agron. J.* 63:593–596.

Blevins, R.L., G.W. Thomas, and P.L. Cornelius. 1977. Influence of no-tillage and nitrogen fertilization on certain soil properties after 5 years of continuous corn. *Agron. J.* 69:383–386.

Blevins, E.L., G.W. Thomas, M.S. Smith, W.W. Frye, and P.L. Cornelius. 1983. Changes in soil properties after 10 years continuous non-tilled and conventionally tilled corn. *Soil Tillage Res.* 3:135–146.

Bond, J.J., and W.O. Willis. 1969. Soil water evaporation: Surface residue rate and placement effects. *Soil Sci. Soc. Am. Proc.* 33:445–448.

Box, J.E., S.R. Wilkerson, R.N. Dawson, and J. Kozachyn. 1980. Soil water effects on no-till corn production in strip and completely killed mulches. *Agron. J.* 72: 797–802.

Breman, J.W., and D.L. Wright. 1984. Using winter legume mulches as a nitrogen source for no-tillage corn and grain sorghum production. In: J.T. Touchton (ed.), *Seventh Annual Southeast No-Tillage Systems Conference*, pp. 6–17. Alabama Agricultural Experimental Station, Auburn University, AL.

Brill, G.D., C.S. Slater, and R.V.D. Broach. 1963. Conservation methods for soils of the Northern Coastal Plan. *U.S.D.A. Agric. Information Bull. 271.*

Brown, P.L., and D.D. Dickey, 1970. Losses of wheat straw residue under simulated field conditions. *Soil Sci. Soc. Am. Proc.* 34:118–121.

Brown, S.M., T. Whitwell, J.T. Touchton, and C.H. Burmester. 1985. Conservation tillage systems for cotton production. *Soil Sci. Soc. Am. J.* 49:1256–1260.

Browning, G.M. 1937. Changes in the erodibility of soils brought about by the application of organic matter. *Soil Sci. Soc. Am. Proc.* 2:85–96.

Buntley, G.J. 1986. Tennessee No-Tillage update. In: R.E. Phillips (ed.), *Southern Region No-Till Conference, Southern Region Series Bulletin 319*, Lexington, KY. pp. 100–102.

Chalk, P.M. 1985. Review: Estimation of N fixation by isotope dilution: An appraisal of techniques involving $^{15}N^2$ enrichment and their application. *Soil Biol. Biochem.* 17:389–410.

Chepil, W.S. 1955. Factors that influence clod structure and erodibility of soil by wind: V. Organic matter at various stages of decomposition. *Soil Sci.* 80:413–421.

Coleman, R. 1941. The value of legumes for soil improvement. *Mississippi Agricultural Experimental Station Bulletin 336.*

Dabney, S.M., R.W. Taylor, and A.W. Nipper. 1984. Evaluation of cover crops for no-till corn silage production in Louisiana. *Proceedings of the Seventh Annual Southeast No-tillage Systems Conference*, pp. 18–21. Alabama Agricultural Experimental Station, Auburn University.

Dabney, S.A., and D.R. Bouldin. 1985. Fluxes of ammonia over an alfalfa field. *Agron. J.* 77:572–578.

Danso, S.K.A. 1986. Review: Estimation of N-fixation by isotope dilution: An appraisal of techniques involving $^{15}N^2$ enrichment and their application—Comments. *Soil Biol. Biochem.* 18:243–244.

D'Itri, F.M. (ed.) 1985. *A Systems Approach to Conservation Tillage.* Lewis Publishers, Inc. Chelsea, Michigan.

Donnelly, E.D., and J.T. Cope. 1961. Crimson clover in Alabama. *Auburn University Agricultural Experiment Station Bulletin 335.*

Ebelhar, S.A., W.W. Frye, and R.L. Blevins. 1984. Nitrogen from legume cover crops for no-tillage corn. *Agron. J.* 76:51–55.

Elkins, D., D. Frederking, R. Marashi, and B. McVay. 1983. Living mulch for no-till corn and soybeans. *J. Soil Water Conserv.* 38:431–433.

Elson, J. 1941. A comparison of the effect of fertilizer and manure, organic matter and carbon-nitrogen ratio on water-stable soil aggregates. *Soil Sci. Soc. Am. Proc.* 6:86–90.

Evans, H.H., D.D. Cockran, and M.E. Harwood. 1954. Vegetable crop production as affected by cover crops. *North Carolina Agronomy Research Report 12.*

Eylands, V.J., and R.N. Gallaher. 1984. Nitrogen fertilizer requirements for grain sorghum following winter legumes in conventional and no-tillage systems. In: J.T. Touchton (ed.), *Proceedings of the Seventh Annual Southeast No-Tillage Systems Conferences.* Alabama Agricultural Experimental Station, Auburn University, AL, pp. 70–75.

Flannery, R.L. 1981. Conventional vs. no-tillage corn silage production. *Better Crops* 66:3–6.

Fred, E.B., I.L. Baldwin, and E.M. McCoy. 1932. Root nodule bacteria and leguminous plants. *Wisconsin University Studies in Science*, 5.

Fredrickson, J.K., F.E. Koehler, and H.H. Cheng. 1982. Availability of 15N-labeled nitrogen in fertilizer and in wheat straw to wheat in tilled and no-till soil. *Soil Sci. Soc. Am. J.* 46:1218–1222.

Fribourg, H.A., and W.V. Bartholomew. 1956. Availability of nitrogen from crop residues during the first and second season after application. *Soil Sci. Soc. Am. Proc.* 20:505–508.

Frye, W.W. 1986. Economics of legume cover crops. *Soil Sciences News and Views*, Department of Agronomy, University of Kentucky, Lexington.

Frye, W.W., W.G. Smith, and R.J. Williams. 1985. Economics of winter cover crops as a source of nitrogen for no-till corn. *J. Soil Water Conserv.* 40:246–258.

Gantzer, C.J., and G.R. Blake. 1978. Physical characteristics of LeSuer clay loam soil following to-till and conventional tillage. *Agron. J.* 70:853–857.

Ghuman, B.S. and R. Lal. 1982. Temperature regime of a tropical soil in relation to surface condition and air temperature and its Fourier analysis. *Soil Sci.* 134:133–140.

Hale, J.K., N.L. Hartwig, and L.D. Hoffman. 1984. Cyanizine losses in rufoff from no-tillage corn in "living" and dead mulches vs. unmulched conventional tillage. *J. Environ. Qual.* 13:105–110.

Hardy, R.W.F., R.C. Burns, and R.D. Holsten. 1973. Applications of the acetylene-ethylene assay for the measurement of nitrogen fixation. *Soil Biol. Biochem.* 5:47–82.

Hargrove, W.L. 1981. Proceedings of the minisymposium on legume cover crops for conservation tillage production systems. *University of Georgia—Atlanta Special Publication 19.*

Hargrove, W.L., G.W. Langdale, and A.W. Thomas. 1984. Role of legume cover crops in conservation tillage production systems. *Am. Soc. Agric. Eng.* Paper 84-2038.

Hargrove, W.L. 1986. Winter legumes as a nitrogen source for no-till grain sorghum. *Agron. J.* 78:70–74.

Hartwig, N.L. 1974. Crownvetch makes a good sod for no-till corn. *Crops Soils* 27:16–17.

Hendrickson, B.H., A.P. Barnett, and O.W. Beale. 1963. Conservation methods for soils of the Southern Piedmont. *U.S.D.A. Agricultural Information Bulletin 269.*

Herridge, D.F., and J.S. Pate. 1977. Utilization of net photosynthate for nitrogen fixation and protein production in an annual legume. *Plant Physiol.* 60:759–764.

Hovis, L. 1943. Aggregation of an orchard and a vegetable soil under different cultural treatments. *Ohio Agricultural Experimental Station Bulletin 640.*

Hoyt, G.D., and W.L. Hargrove. 1986. Legume cover crops for improving crop and soil management in the Southern United States. *Hort Science* 21:397–402.

Huntington, T.G., J.H. Grove, and W.W. Frye. 1985. Release and recovery of nitrogen from winter annual cover crops in no-till corn production. *Commun. Soil Sci. Plant Anal.* 16:193–211.

Jensen, H.L., and D. Frith. 1944. Production of nitrate from root and root nodules of legume and subterranean clover. *Proc. Linn. Soc. N.S.W.* 69:210–214.

Jones, R.J. 1942. Nitrogen losses from Alabama soils in lysimeters as influenced by various systems of green manure crop management. *Agron. J.* 34:574–585.

Jordan, H.V., S.P. Crockett, and C.E. Bardsley, Jr. 1956. Some effects of kudzu versus continuous corn or soil properties and crop yields. *Soil Sci. Soc. Am. Proc.* 20: 225–227.

Juaregui, H.E.M. 1986. *Management of Perennial and Reseeding Legume Cover Crops for No-Tillage Corn Production.* M.S. thesis, University of Kentucky, Lexington.

Kamprath, E.J., W.V. Chandler, and B.A. Krantz. 1958. Winter cover crops: Their effects on corn yields and soil properties. *North Carolina Agricultural Experimental Station Technical Bulletin 129.*

Karraker, P.E., C.E. Bortner, and E.N. Fergus. 1950. Nitrogen balance in lysimeters as affected by growing Kentucky bluegrass and certain legumes separately and together. Kentucky Agricultural Experimental Station Bulletin 557.

Kitur, B.K. 1982. *Fate of nitrogen in no-tillage and conventional tillage corn systems using labeled NH_4NO_3.* M.S. thesis, University of Kentucky, Lexington.

Knavel, D.E., and J.W. Herron. 1981. Influence of tillage system, plant spacing and nitrogen on head weight, yield, and nutrient concentration of spring cabbage. *J. Am. Soc. Hort. Sci.* 106:540–545.

Kumar Rao, J.V.D.K., P.J. Dart, and P.V.S.S. Sastry. 1983. Residual effect of pigeonpea (*Cajanus cajan*) on yield and nitrogen response of maize. *Exper. Agric.* 19:131–141.

Kuo, M.H., and W.V. Bartholomew. 1963. On the genesis of organic nitrogen in decomposed plant residue. In: *The Use of Isotopes in Soil Organic Matter Studies.* FAO/IAEA.

Ladd, J.N., J.M. Oades, and M. Amato. 1981. Microbial biomass formed from ^{14}C, ^{15}N-labelled plant material decomposing in soils in the field. *Soil Biol. Biochem.* 13:119–126.

Ladd, J.N., M. Amato, R.B. Jackson, and J.H.A. Butler. 1983. Utilization by wheat crops of nitrogen from legume residues decomposing in soils in the field. *Soil Biol. Biochem.* 15:231–238.

Ladd, J.N., and M. Amato. 1986. The fate of nitrogen from legume and fertilizer sources in soils successively cropped with wheat under field conditions. *Soil Biol. Biochem.* 18:417–425.

Lal, R. 1974. No-tillage effects on soil properties and maize (*Zea mays* L.) production in Western Nigeria. *Plant Soil* 40:321–331.

Lal, R. 1976. No-tillage effects on soil properties under different crops in Western Nigeria. *Soil Sci. Soc. Am. Proc.* 40:762–768.

Lal, R., D. DeVleeschauwer, and R. Malafa Nganje. 1980. Changes in properties of a newly cleared tropical alfisol as affected by mulching. *Soil Sci. Soc. Am. J.* 44: 827–833.

Langdale, G.W., A.P. Barnett, R.A. Leonard, and W.G. Fleming. 1979. Reduction of erosion by the no-till system in the Southern Piedmont. *Trans. Amer. Soc. Agr. Eng.* 22:82–92.

Larson, W.E., C.E. Clapp, W.H. Pierre, and Y.B. Morachan. 1972. Effects of increasing amounts of organic residues on continuous corn: II. Organic carbon, nitrogen, phosphorus and sulfur. *Agron. J.* 64:204–208.

Lohnis, F. 1926. Nitrogen availability of green manures. *Soil Sci.* 22:253–290.

Lyon, T.L. 1936. The residual effects of some leguminous crops. *New York* (Cornell) Agricultural Experimental Station Bulletin 645.

Mannering, J.V., and L.D. Meyer. 1963. The effects of various rates of surface mulch on infiltration and erosion. *Soil Sci. Soc. Am. Proc.* 27:84–86.

Martin, G.W., and J.T. Touchton. 1982. Management of winter legumes for grain sorghum production. *Abstracts of Technical Papers*, No. 9. Southern Branch of the American Society of Agronomy, Orlando, FL.

Materon, L., and C. Hagedorn. 1982. Nodulation of crimson clover by introduced rhizobia in Mississippi soils. *Soil Sci. Soc. Am. J.* 46:553–556.

McCalla, T.M. 1945. Influence of microorganisms and some organic substances on soil structure. *Soil Sci.* 59:287–297.

McCalla, T.M. 1959. Microorganisms and their activity with crop residues. Nebraska Agricultural Experimental Station Bulletin 453.

McCalla, T.M., and F.L. Duley. 1943. Distintegration of crop residues as influenced by subtillage and plowing. *Agron. J.* 35:306–315.

McCalla, T.M., and F.L. Duley. 1948. Stubble mulch studies: Effect of sweetclover extract on corn germination. *Science* 108:163.

McCalla, T.M., and J.C. Russel. 1948. Nitrate production as affected by sweetclover residues left on the surface of the soil. *Agron. J.* 40:411–421.

Megie, C.A., R.W. Pearson, and A.E. Hiltbold. 1967. Toxicity of decomposing crop residues to cotton germination and seedling growth. *Agron. J.* 59:197–199.

Meyer, L.D., W.H. Wischmeier, and G.R. Foster. 1970. Mulch rates required for erosion control on steep slopes. *Soil Sci. Soc. Am. Proc.* 34:928–931.

Meyers, R.J.K., and E.A. Paul. 1971. Plant uptake and immobilization of ^{15}N labeled ammonium nitrate in a field experiment with wheat. In: *Nitrogen 15 in soil-plant studies.* IAEA, Vienna.

Mitchell, W.H., and M.R. Teel. 1977. Winter annual cover crops for no-tillage corn production. *Agron. J.* 69:569–573.

Moschler, W.W., G.M. Shear, D.L. Hallock, R.D. Sears, and G.D. Jones. 1967. Winter cover crops for sod-planted corn: Their selection and management. *Agron. J.* 59:547–551.

Nelson, M. 1944. Effect of the use of winter legumes on yields of cotton, corn and rice. *Arkansas Agricultural Experimental Station Bulletin 451.*

Nelson, L.R., R.N. Gallaher, R.R. Bruce, and M.R. Holmes. 1977. Production of corn and sorghum grain in double cropping systems. *Agron. J.* 69:41–45.

Norman, A.G., and C.H. Werkman. 1943. The use of the nitrogen isotope N15 in determining nitrogen recovery from plant material decomposing in soil. *Agron. J.* 35:1023–1025.

Obenshain, S.S., and P.T. Gish. 1941. The effect of green manure crops on certain properties of Berks silt loam. *Virginia Agricultural Experimental Station Technical Bulletin 73.*

Palada, M.C., S. Ganser, R. Hofstetter, B. Volak, and M. Culik. 1983. Association of interseeded legume cover crops and annual row crops in year-round cropping systems. In: W. Lockeretz (ed.) *Environmentally Sound Agriculture*, Praeger Scientific, New York, NY.

Parker, D.T. 1962. Decomposition in the field of buried and surface applied cornstalk residue. *Soil Sci. Soc. Am. Proc.* 28:559–562.

Phillips, R.E. 1984. Soil moisture. In: R.E. Phillips and S.H. Phillips (eds.), *No-Tillage Agriculture: Principles and Practice*. VanNostrand Reinhold Co., New York.

Phillips, R.E., and S.H. Phillips (eds.). 1984. *No-Tillage Agriculture: Principals and Practices*. VanNostrand Reinhold Co., New York.

Pieters, A.J. 1927. *Green Manuring Principles and Practice*. Wiley, New York.

Pieters, A.J., and R. McKee. 1929. Green manuring and its application to agricultural practices. *Agron. J.* 21:985–993.

Pieters, A.J., P.R. Henson, W.E. Adams and A.P. Barnett. 1950. Sericea and other perennial lespedezas for forage and soil conservation. *U.S.D.A. Circular 863*.

Pieters, A.J., and R. McKee. 1938. The use of cover and green manure crops. In: *Soils and Men. U.S.D.A. Yearbook of Agriculture*.

Reeves, D.W., D.H. Rickerl, C.B. Elkins, and J.T. Touchton. 1986. No-tillage update report—Alabama. In: R.E. Phillips (ed.), *Southern Region No-Till Conference, Southern Region Series Bulletin 319*, Lexington, KY.

Rogers, T.H., and J.E. Giddens. 1957. Green manuring and cover crops. In: A. Stefferund (ed.), *Soils. The Yearbook of Agriculture*. U.S.D.A., pp. 252–257.

Russell, E.J. 1966. *A History of Agricultural Science in Great Britain, 1620–1954*. George Allen and Unwin, Ltd., London.

Saloman, M. 1962. Soil aggregation-organic matter relationship in redtop-potato rotations. *Soil Sci. Soc. Am. Proc.* 26:51–54.

Semple, E.C. 1928. Ancient Mediterranean Agriculture: Part II. Manuring and seed selection. *Agric. Hist.* 2:129–156.

Siddoway, F.H. 1963. Effects of cropping and tillage methods on dry aggregate soil structure. *Soil Sci. Soc. Am. Proc.* 27:452–454.

Smith, G.R., and W.E. Knight. 1984a. Inheritance of ineffective nodulation in crimson clover. *Crop Sci.* 24:601–604.

Smith, G.R., and W.E. Knight, 1984b. Breeding legumes for improved N_2-fixation. In: G.B. Collins and J.G. Petolino (eds.), *Applications of Genetic Engineering to Crop Improvement* pp. 1–23 Martinus Nijhoff: Dordrecht, The Netherlands.

Smith, G.R., W.E. Knight, and H.L. Peterson. 1982. The inheritance of N_2 fixation efficiency in crimson clover. *Crop Sci.* 22:1091–1094.

Smith, G.R., W.E. Knight, H.L. Peterson, and C. Hagedorn. 1982. The effect of *Rhizobium trifoli* strains and crimson clover genotypes on N_2 fixation. *Crop Sci.* 22:970–972.

Stickler, F.C., W.D. Shrader, and I.J. Johnson. 1959. Comparative value of legume and fertilizer nitrogen for corn production. *Agron. J.* 51:157–160.

Strickling, E. 1950. The effect of soybeans on volume weight and water stability of soil aggregates, soil organic matter content and crop yield. *Soil Sci. Soc. Am. Proc.* 15:30–34.

Stott, D.E., G. Kassim, W.M. Jarrell, J.P. Martin and K. Haider. 1983. Stabilization and incorporation into biomass of specific plant carbons during biodegration in soil. *Plant Soil* 70:15–26.

Thomas, G.W., R.L. Blevins, R.E. Phillips, and M.A. McMahon. 1973. Effect of a killed sod mulch on nitrate movement and corn yield. *Agron. J.* 65:736–739.

Touchton, J.T., W.A. Gardner, W.L. Hargrove, and R.R. Duncan. 1982. Reseeding crimson clover as a N source for no-tillage grain sorghum production. *Agron. J.* 74:283–287.

Touchton, J.T., D.H. Rickerl, R.H. Walker, and C.E. Snipes. 1984. Winter legumes as a nitrogen source for no-tillage cotton. *Soil Tillage Res.* 4:391–401.

Triplett, G.B., F. Haghiri, and D.M. Van Doren. 1979. Plowing effect of corn yield response to N following alfalfa. *Agron. J.* 71:801–803.

Triplett, G.B., Jr., D.M. Van Doren, Jr., and B.L. Schmidt. 1968. Effect of corn stover mulch on no-tillage corn yield and water infiltration. *Agron. J.* 60:236–239.

Uhland, R.E. 1949. Physical properties of soils as modified by crops and management. *Soil Sci. Soc. Am. Proc.* 14:361–366.

Unger, P.W. 1978. Straw-mulch rate effect on soil water storage and sorghum yield. *Soil Sci. Soc. Am. J.* 42:486–491.

U.S. Department of Agriculture, Soil Conservation Service. 1978. *Predicting Soil Loss Using the Universal Soil Loss Equation.* Lexington, KY.

Utomo, M. 1986. *Role of Legume Cover Crops in No-Tillage and Conventional Tillage Corn Production*, Ph.D. thesis, University of Kentucky, Lexington.

Vallis, I., and R.J. Jones. 1973. Net mineralization of nitrogen in leaves and leaf litter of *Desmodium intortum* and *Phaseolus atropurpureus* mixed with soil. *Soil Biol. Biochem.* 5:391–398.

Van Doren, C.A., and R.S. Stauffer. 1943. Effect of crop and surface mulches on runoff, soil losses, and soil aggregation. *Soil Sci. Soc. Am. Proc.* 8:97–101.

Van Wijk, W.R., W.E. Larson, and W.C. Burrows. 1959. Soil temperature and the early growth of corn from mulched and unmulched soil. *Soil Sci. Soc. Am. Proc.* 23:428–434.

Varco, J.J. 1986. *Tillage Effects on Transformation of Legume and FertilizerNitrogen and Crop Recovery of Residue Nitrogen.* PhD. thesis, University of Kentucky, Lexington.

Varco, J.J., W.W. Frye, J.H. Herbek, and R.L. Blevins. 1984. Effects of legume cover crops on yield of no-till corn. In: J.T. Touchton (ed.), Proceedings of the Seventh Annual Southeast No-Tillage Systems Conference, Alabama Agricultural Experimental Station, Auburn University AL.

Vrabel, T.E., P.L. Minotti, and R.D. Sweet. 1981. Legume sods as living mulches in sweet corn. *Proc. N.E. Weed Sci. Soc.* 35:158–159.

Wade, M.K., and P.A. Sanchez. 1983. Mulching and green manure applications for continuous crop production in the Amazon Basin. *Agron. J.* 75:39–45.

Wagger, M.G., D.E. Kissel, and S.J. Smith. 1985. Mineralization of nitrogen from nitrogen-15 labeled crop residue and under field conditions. *Soil Sci. Soc. Am. J.* 49:1220–1226.

Waksman, S.A., and R.L. Starkey. 1931. *The Soil and the Microbe.* Wiley, New York.

Wedderbuan, C.H.M., and W.G. Collingwood. 1976. *The Economist of Xenophon.* Lenox Hill Publishing and Distributing Co., New York.

Welch, C.D., W.L. Nelson, and B.A. Krantz. 1950. Effects of winter cover crops on soil properties and yields in a cotton-corn and in a cotton-peanut rotation. *Soil Sci. Soc. Am. Proc.* 15:229–234.

Wilkinson, S.R., and J.W. Dobson, Jr. 1981. Unpublished data. Department of Agronomy, Georgia Agricultural Experimental Station, Athens.

Wilson, D.O., and W.L. Hargrove. 1986. Release of nitrogen from crimson clover residue under two tillage systems. *Soil Sci. Soc. Am. J.* 40:1251–1254.

Wilson, G.F., R. Lal, and B.N. Okigbo. 1982. Effects of cover crops on soil structure and on yield of subsequent arable crops grown under strip tillage on an eroded Alfisol. *Soil Tillage Res.* 2:233–250.

Wisniewski, A.J., M. Soloman, and J.B. Smith. 1958. Restoration of soil structure with redtop and other sod crops. *Soil Sci. Soc. Am. Proc.* 22:320–322.

Witt, W.W. 1984. Response of weeds and herbicides under no-tillage conditions. In: R. E. Phillips and S.H. Phillips (eds.), *No-Tillage Agriculture: Principles and Practices.* VanNostrand Reinhold Com, New York.

Worsham, A.D. 1986. No-tillage research update—North Carolina. In: R.E. Phillips (ed.), *Southern Region No-Till Conference, Southern Region Series Bulletin 319*, Lexington, KY.

Yaacob, O., and G. Blair. 1980. Mineralization of ^{15}N-labelled legume residues in soils with different nitrogen contents and its uptake by Rhodes grass. *Plant Soil* 57:237–248.

Yogeswara Rao, Y., and W. Ali (1983). Grain and fodder legumes as preceeding crops affecting the yield and N economy of rice. *J. Agric Sci.* 101:463–466.

Influence of Sludge Organic Matter on Soil Physical Properties

L. Metzger and B. Yaron*

I. Introduction

In the last two decades, land disposal of sludge expanded after it appeared to be a generally advantageous and environmentally sound method. In addition to its fertilizing value, this organic material is likely to improve the soil physical condition. The gradual decrease in the organic matter content in cultivated soils, which may lead to deterioration of the soil physical status

*Division of Soil Organic and Residues Chemistry, Institute of Soils and Water, The Volcani Center, Bet Dagan 50250, Israel. Contribution from the Agricultural Research Organization, The Volcani Center, Bet Dagan, Israel. No. 1744-E, 1986 series.

and erosion, is particularly worrying in warm areas, where organic matter decomposes rapidly. The application of organic wastes such as sludge has gained renewed interest as a means of maintaining sufficiently high organic matter contents in soils.

Among the multitude of publications concerned with land disposal of sludge, only a small percentage is devoted to the effect of sludge on soil physical properties. Guidi (1981), in his review, indicated clearly which changes in the soil physical properties occurred after sludge application. He did not, however, provide comprehensive information on the chemical and biological processes involved in the modification of the soil physical properties. Yaron et al. (1984) gathered some information on the effect of solid organic components of sewage on the hydraulic status of soils. Summarizing the results published on the effect of sludge and other organic wastes on soil physical properties, Khaleel et al. (1981) established correlations between the changes in the properties reported and the increase in the organic carbon (OC) content of the soil. This correlation study, which is helpful for analyzing the results of the effect of sludge on water-holding capacity and bulk density, is not applicable when modifications of the aggregation status and the water transmission properties are considered. Complementary reviews on the effect of green manures (McRae and Mehuys, 1985) and soil microorganisms (Lynch and Bragg, 1985) on soil physical properties provide relevant information on the factors affecting and the mechanisms involved in the modification of the soil physical status due to addition of sludge.

The introduction of sludge into the soil system promotes the development of chemical reactions (redox reactions, hydrolysis, etc.), physicochemical reactions (dissolution, precipitation, adsorption, bonding, etc.), and microbiological processes (microbial multiplication, metabolite production, enzymatic reactions, etc.). Organic substances released in the medium, regardless of their origin, are involved in various interactions with the mineral fraction of the soil. Studying these interactions and their evolution with time is essential for an understanding of the modifications of the soil physical properties after sludge additions. Thus, in the first part of this chapter, the existing knowledge on sludge organic compounds-soil interactions is summarized. In the second part, the modifications of the physical properties brought about by sludge are reported and tentatively explained by the previously mentioned interactions. In each part of this chapter, the results published in the literature are examined and discussed in the light of the authors' own findings.

II. Sludge Organic Matter Interactions with Soils

A. Persistence of Sludge in Soil Medium

The persistence of the effects of sludge on soil physical properties is directly related to the persistence of sludge OC in the soil medium. If part of the sludge OC is resistant to degradation, the OC content of the soil will increase under repeated sludge applications and an irreversible alteration of the physical properties is possible. The organic decay is commonly assessed by determining the OC evolved as CO_2 or the OC remaining in the soil after wastes application, or through the organic N mineralization. In most cases the potential of sludge organic matter mineralization was approximated from results obtained during incubation of soil–sludge mixtures under constant conditions.

1. Variations in Potential of Mineralization for Different Sludges

Percentages of sludge OC mineralized after long periods of residence in soil, as reported in the literature, are presented in Table 1. The extent of mineralization is strongly affected by the type of sludge utilized or, more specifically, the type of treatment to which sludge is subjected before application. For

Table 1. Percentage of sludge organic carbon mineralized after various periods of residence in soils

Reference	Type of sludge	Duration of contact	%OC mineralized[a]
Field conditions			
Hohla et al., 1978	Liquid anaerobically digested	6 years[b]	60
Gupta et al., 1977	Dry anaerobically digested	1 year 2 years	42 50
Hall and Coker, 1983	Liquid undigested	3 years	77–82
Laboratory conditions[c]			
Miller, 1974 (var.)	Dry anaerobically digested	6 months	15–20
Terry et al., 1979a (21°C)	Dry anaerobically digested	4 months 11 months	26–42 39–46
Terry et al., 1979b (21°C)	Dry ^{14}C-labeled synthetic	11 months	32–46
Tester et al., 1977 (n.s.)	Undigested composted	2 months	16
Agbim et al., 1977 (22°C)	Dry anaerobically digested	12 months	19–29
Morel and Jacquin, 1976 (28°C)	Dry anaerobically digested, flocculated	6 months	21
	Dry anaerobically disgested	6 months	23
	Various	1 month	9–29
Metzger, 1986 (35°C)	Dry undigested	6 months 8 months	57 64

[a] Average or range.

[b] Sludge disposed in equal yearly applications throughout the experimental period.

[c] The temperatures of incubation are given in parentheses: var. = variable; n.s. = not specified.

anaerobically digested sludges, added to different soils and incubated, only about 20% of the OC degraded during 6 months (Miller, 1974; Morel and Jacquin, 1976). For similar sludges, Terry et al. (1979a) found a wider range of mineralization (26–42%) after 4 months of incubation. A relatively high percentage of OC resistant to degradation was also observed in field experiments conducted with the same type of sludge: 42% of the OC from such a sludge degraded during 1 year after application to a sandy soil. Whereas the extent of OC mineralization is similar for anaerobically digested and composted sludges (Table 1), the amounts of mineralizable OC are much higher in the case of undigested sludges. After the incubation of dry undigested sludge within sandy and loamy soils for a period of 7–9 months, we observed that at least 60% of the sludge OC had been mineralized (Metzger, 1986). This finding confirms the results of Hall and Coker (1983), who reported ~80% OC mineralization 3 years after the application of liquid undigested sludge to grassland. It may, however, be observed that a large part of any type of sludge OC introduced into the soil medium is either resistant to microbial oxidation or rapidly converted into stable forms of humic materials. Consequently, repeated application of sludge in the field will lead to an increase in the OC content of the soil. This was confirmed by the experiment of Hohla et al. (1978), in which altogether 298 t/ha of sludge was applied to a silt loam, over the course of 6 years. The OC budget showed that ~60% of the sludge OC was lost during this period, whereas the OC content of the surface horizon increased from 0.95 to 2.29%.

In the evaluation of the persistence of sludge organic matter in soils, results deduced only from the amounts of organic N mineralized may be misleading. In this case, the variability of the results obtained is much greater than this observed for the measurements of the OC mineralization. Summarizing the existing information on N mineralization in sludge-amended soils, Fine et al. (1986) reported N mineralization values ranging from 1 to 70% for various sludges, conditions, and authors. In addition, the fate of the applied organic N does not always reflect the fate of the major part of the organic matter, since under certain conditions sludge mineralization may even occur concurrently with N immobilization. The C:N ratio of the organic waste applied affects markedly the extent of N mineralization. Microorganisms use the C and N found in the freshly added material to build new living tissues. The C:N ratio of microflora cells averages approximately 10. During microbial degradation of a waste with a much higher C:N, mineral N found in the soil will be immobilized and transformed into organic N and global mineralization may be retarded (Alexander, 1977). Incorporation of a waste with a much smaller C:N ratio will lead to liberation of mineral N in excess. When a compost obtained from a limed sludge and woodchips mixture (C:N = 17) was incorporated into different types of soil for 54 days, ~16% of the compost OC evolved as CO_2, whereas apparently only 6% of the compost N was mineralized. This was explained as being due to N immobilization or ammonium fixation in clays (Tester et al., 1977). In contrast, when anaerobically digested sludge with a C:N ratio of 7.8 was applied to a silt loam in the field, organic N was lost at a faster rate than OC (Hohla et al., 1978). As a consequence of the confusion inherent in the organic N mineralization reports, organic matter mineralization should be evaluated from publications concerned with the OC degradation.

2. Kinetics of Sludge OC Mineralization

The pattern of OC decay after sludge application is characterized by a rapid decrease followed by progressive stabilization as the ordinate seems to approach a hypothetical "equilibrium value." Examples of this pattern are given in Figure 1, showing the OC mineralization of an undigested sludge mixed with a loamy soil, and in Figure 2A (Curve 2), showing the OC mineralization of the same sludge mixed with a sandy soil (Metzger, 1986). The overall decomposition of sludge OC can be described schematically as a stepwise series of steady state reactions corresponding to the decomposition of chemical fractions of different degradabilities. For each fraction, the C decay can be mathematically described by a first-order kinetic equation (Gilmour et al., 1977):

$$C(t) = C_0 e^{-k(t-t_0)} \quad [1]$$

where $C(t)$ is residual OC amount at time t, C_0 is initial OC input (this fraction),

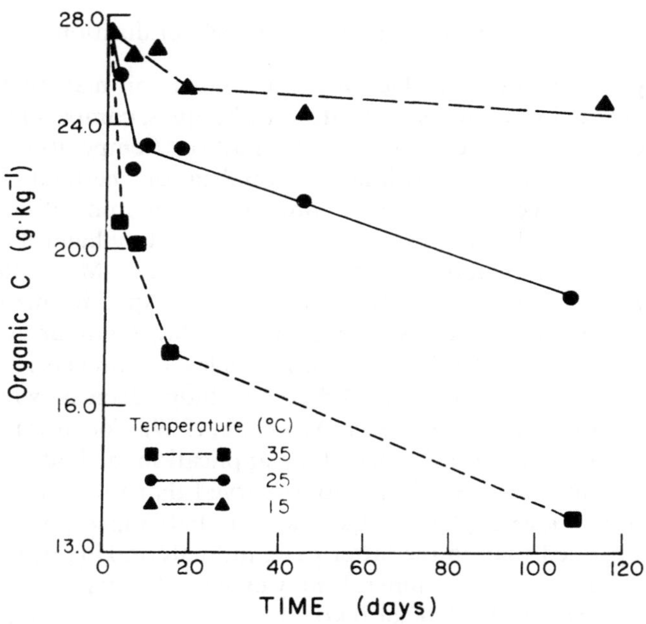

Figure 1. Effect of temperature on the kinetics of organic carbon mineralization in a loamy soil–sludge (5%) mixture.

k is mineralization rate constant, and t_0 is time at which the degradation of this fraction begins. During each phase, k can be calculated as:

$$k = \frac{2.303}{t_2 - t_1} \log C_1/C_2 \qquad [2]$$

Organic C decay curves obtained from experimental results are usually divided into two or three successive phases. The OC decomposition in soil of a synthetic sludge (21°C) begins with a rapid phase lasting 28 days, followed by a second, slower phase until the remainder of the OC is completely decomposed (Terry et al., 1979a). Studying the biodegradation of various sludges at 28°C, Morel and Jacquin (1977) distinguished an initial phase in which the easily available substrates are utilized by the microorganisms (0–5 days), an intermediary phase of decreasing biological activity (5–14 days), and a third phase during which the resistant OC is slowly mineralized. Dealing with the decomposition of an undigested sludge in a loamy soil at 15 and 25°C, we found that the mineralization process may be divided into two phases, represented by two linear segments in the graph log(OC) versus time (Figure 1); yet, at a higher temperature (35°C), the distinction of three phases was preferable. Under the assumption that the first-order kinetic model is applicable throughout the incubation, a different mineralization rate constant k can be calculated for each phase (Gilmour et al., 1977; Gilmour and Gilmour, 1980).

3. Effect of Various Parameters on Sludge OC Mineralization

The decomposition rate of sludge OC in amended soil may be affected by sludge characteristics (source, stabilization techniques, composition), soil parameters (e.g., texture, pH), environmental conditions (temperature, moisture), and application conditions (loading rate, application method). The strong influence of the sludge stabilization techniques on the mineralization rate is demonstrated above (Table 1, section II.A.1). On the other hand, it is generally reported that soil parameters have little effect on k values (Miller, 1974; Tester et al., 1977; Sommers et al., 1979; Terry et al., 1979b). Applying manure to soil at loading rates varying between 45 and 225 t/ha, Gilmour et al. (1977) observed nearly identical k values. In contrast, a few researchers reported that annual decomposition percentages of digested sludge decrease when loading rates increase (Agbim et al., 1977; Sommers et al., 1979). Too little information is available to ascertain the influence of the application method on the sludge mineralization rate. Terry et al. (1979b) observed that OC mineralization is more rapid for surface-applied sludge than for sludge incorporated into the soil. They suggested that, in the latter case, intimate mixing between sludge organic constituents and clay minerals may explain the retardation of the OC mineralization. In the field, Kladivko and Nelson (1979a) compared the dynamics of the OC content of a silt loam soil, to which liquid sludge was applied by three different methods (surface application, disking-in, and rototilling). Their data indicate that disked-in sludge decomposed more slowly than both surface-applied and rototilled sludge, but the differences between the treatments were not significant.

The dominant factors that determine microbial activity, and hence the sludge mineralization rate, are moisture and temperature. Soil moisture affects the decomposition of OC from waste origin in two ways. In the low moisture range, the water supply for the growth of microorganisms is insufficient. In the high moisture range, the oxygen supply to microorganisms becomes deficient. In between there is an optimum moisture content for microbial activity; close to this optimum there is little effect of moisture change upon the mineralization rate constant (k), and at both higher and lower moisture contents k decreases significantly (Gilmour and Gilmour, 1980). The optimal water content usually corresponds to "field capacity." In the water potential range -0.25 to -1 bar, k for sludge OC is nearly independent of water content (Terry et al., 1979b).

The mineralization of any organic material in the soil is most rapid at an optimum temperature. This optimum varies with the composition of the soil microbial population. For a mesophilic microflora, maximum microbial activity and, hence, maximum rates of OC decay take place at 30–35°C. In OC decomposition studies, most of which were carried out at temperatures in the range of 5–35°C, any increase in temperature accelerates the decomposition rate (Reddy et al., 1980). Accordingly, Miller (1974), who reproduced the effect of outdoor temperature variations upon sludge degradation, observed that the extent of OC mineralization after a given time is correlated with the heat input. To express the magnitude at which the decomposition process depends on temperature, the temperature coefficient Q_{10} is commonly used. Q_{10} is defined as the ratio between mineralization rates (k) at two temperatures separated by a 10°C interval. In practice, k is clearly related to temperature, hence, Q_{10} has a meaningful value only for the initial rapid mineralization phase. Terry et al. (1979b) reported that k of an anaerobically digested sludge in soil was nearly independent of temperature from 56 days to the end of the incubation. They found that during the phase of intense biological activity (0–12 days), Q_{10} is ~ 1.7 for the range of temperatures 15–35°C. In an experiment carried out in our laboratory on the OC mineralization for an undigested dry sludge incorporated in a loamy soil, it was found that under increasing temperature the duration of the rapid-mineralization phase decreased (Figure 1). By using the average mineralization rates calculated for this shorter or longer phase, we obtained Q_{10} values of 5.6 and 2.9 for the 15–25°C and 25–35°C temperature ranges, respectively. The latter Q_{10} value, like that reported by Terry et al. (1979b), corresponds roughly to the temperature coefficient characteristic of enzymatic biological reactions. For prediction purposes, a standard Q_{10} of 2.0 is often assumed (Gilmour et al., 1977).

B. Effect of Sludge on Composition of Soil Organic Matter

Scant and contradictory information is available on the changes in the composition of soil organic matter following sludge application. The conflicting results can be explained in part by distinguishing between data obtained after short periods of incubation and those obtained after longer periods. Since our objective is to understand how sludge–soil organic matter interactions affect the soil physical status, the organic fractions of main interest here are the hydrophobic compounds and the polysaccharides.

By comparing the distribution of several fractions of the soil OC before and after amendment with sludge, Hohla et al. (1978) found a sevenfold increase in the percentage of oil and grease in the OC at the end of a 6-year experiment. Still, the anaerobically digested sludge used contained a remarkably high and unusual level of this hydrophobic fraction. The accumulation of this fatty fraction may induce the formation of hydrophobic aggregates, thereby affecting the soil physical properties. The results of OC gross fractionation analyses carried out by Hiroux et al. (1983) on limed sludge–soil mixtures incubated for 4 months disagree with the above-reported results: the ether-soluble fraction (oils, fats, waxes) of the sludge OC was mineralized at a rapid rate from the onset of incubation and a minor residue of it remained in the soil after 9 weeks. According to these workers, the water-soluble and the nitrogenous "humus-lignin" fractions of the soil OC increased as a result of the sludge application. Metzger (1986) investigated the effect of a dry undigested sludge on the ether-soluble and water-soluble fractions of the OC of a loamy soil after 256 days of incubation at 35°C. During this period of incubation, the ether-soluble fraction dropped from 7.9 to 5.8% of the total OC, while the water-soluble fraction rose from 7.6 to 9.9%. These results, in close agreement with those of Hiroux et al. (1983) quoted previously, support the view that the sludge organic compounds soluble in ether are rapidly decomposable in soils. The changes in the fractionation results over several months of incubation, during which the main part of the sludge OC degraded, show also that there is a progressive adjustment of the OC composition in the amended soil to that in the native soil. These results suggest that on a long-term scale, the organic matter composition in a sludge-amended soil would reflect that of sludge organic matter, at a degree increasing with the sludge OC residue: total OC ratio.

Schaumberg et al. (1980) studied the qualitative changes in the composition of the water-soluble fraction of the OC during the incubation of sludge–soil mixtures performed over a period of 100 weeks. From the interpretation of the infrared spectra of the soil extracts collected after different times, they concluded that polysaccharides and compounds of the protein group had been metabolized after 10 weeks of residence in the soil. The importance of the carboxylates group was enhanced at the same time. In our laboratory, we studied the changes in the polysaccharide concentration in silt loam and loamy soil–sludge (5%) mixtures incubated at 25°C for 2 months. The concentrations of two subgroups—the water-soluble polysaccharides (WSP) and the acid-hydrolyzable polysaccharides (AHP)—were estimated. It was observed that the content of these two fractions was immediately enhanced as a result of sludge addition, reflecting the contribution of the polysaccharidic components of the sludge. In a first step of the incubation (9 days), the WSP concentration increased rapidly for both soils amended with sludge; thereafter it stabilized in the silt loam and dropped back in the loamy soil. It is assumed that during the initial period of intensive sludge OC decomposition, the steep increase in the percentage of water-soluble polysaccharides is due to the secretion of polysaccharides by microorganisms.

The usual fractionation scheme in soil humus studies (alkaline extraction) was sometimes used to assess the long-term variations of the soil OC composition after sludge application. The results from laboratory and field experiments reported by Morel et al. (1977) show that the humin fraction was enhanced in sludge-amended soils and that 3 years of disposal led to a decrease in the HA:FA (Humic Acid:Fulvic Acid) ratio. Humic substances extracted from a sludge-amended soil 2 months after incorporation have properties similar to those observed for the sludge humic substances. An identical but attenuated trend persists 12 months later (Boyd et al., 1980). Humic substances formed as a consequence of sludge decay in the soil may be either transformation products of the resistant organic compounds found in sludge, or neoformation products of microbial origin (Terry et al., 1979a). The observations made by Boyd et al. (1980) seem to show that the first process is dominant, but no definite conclusion can be established from the limited research results available so far on sludge humification.

III. Effect of Sludge Organic Matter on Soil Physical Properties

The response of soil to sludge application depends on many factors, such as sludge type and composition, application level and method, soil properties, and climatic conditions. However, the amounts and the composition of the organic compounds incorporated into the soil organic matter following sludge application are undoubtedly the decisive factors explaining the modification of the physical status caused by sludge. The information available, which is examined below, dealt with the effect of sludge or organic C-containing wastes of similar nature on the water retention characteristics, the structure, and the water movement properties of soils.

A. Water Retention Properties

When sludge is applied to soil, the OC input may affect water rentention in two ways. The first and direct effect is due to sludge organic particles themselves, which have a high water adsorption capacity and thus enhance the soil's water retention capacity (Gupta et al., 1977). In an indirect manner, modifications of other physical properties (bulk density, porosity, pore size distribution) due to sludge application affect the water retention properties of the soil.

Generally, the application of sludge raised the water retention capacity of soils. Liquid sludge (30–60 t/ha) increased the volumetric water content of a sandy loam to a depth of 45 cm for as long as 10 and 11 months after sludge application (Kelling et al., 1977). Throughout two cropping seasons after heavy applications of digested and composted sludges (silt loam, 240 t/ha), soil moisture contents remained significantly higher in amended plots (Epstein et al., 1976). Increases in the water retention capacity at field capacity (matric potential (MP) = −0.33 bar) and at wilting point (MP = −15 bar) as a consequence of sludge application are widely reported in the literature (Morel et al., 1978; Kladivko and Nelson, 1979a; Chang et al., 1983; Hall and Coker, 1983; Jacobowitz and Steenhuis, 1984; Wei et al., 1985). The entire water retention curve (water content versus MP) shifted toward higher water contents for a sandy soil sampled at the end of the 2nd year following dry sludge application. The extent of the shift was proportional to the application rate between 225 and 900 t/ha (Gupta et al., 1977). Similarly, the shift of the water retention curve was roughly proportional to the sludge: soil ratio (in the range 1:99 to 10:90) in sandy and loamy soil–sludge mixtures, immediately after mixing (Metzger, 1986). For a silt loam soil mixed with sludge, the shift of the retention curve was greater for raw than for digested sludge (Epstein, 1975). Khaleel et al. (1981) analyzed experimental results obtained in a broad range of soils amended with various organic wastes. They noted that water retention (on a weight basis) increases at both field capacity and wilting point when organic wastes are added, and that relative increases in water retention capacity are greater for coarse-textured soils than for fine-textured soils. Investigating the dynamics of the water retention at field capacity and at wilting point in a sandy soil–sludge (5%) mixture, we observed that the extent of relative increases in water retention at both MPs decreased continuously during a 190-day incubation, but remained significant (22% and 40%, respectively) at the end of this period.

Changes in the available water content (AW) due to sludge application are still a subject of controversy. AW, the soil storage capacity utilizable for plant growth, is usually calculated as the moisture content at −0.33 bar less the moisture content at −15 bar. According to Gupta et al. (1977), for any water potential, most of the increase in water retention capacity due to sludge results from water retained even at −15 bar MP, so AW will be unchanged. The same conclusions may be drawn from the results of Kladivko and Nelson (1979a). Khaleel et al. (1981) claim also that as a result of the bulk density decrease after OC addition, even when increases in AW are evident on a weight basis, they are not significant on a volumetric basis. On the other hand, 9 years of sludge application raised the AW of a sandy loam by 35% (Hall and Coker, 1983). Heavy application of digested sludge and compost in the field led to an increase of the AW after an unspecified time (Epstein et al., 1976). Morel et al. (1978) found that after 3 years of sludge application, AW was significantly increased only for cumulative loading rates higher than 100 t/ha. Our own results suggested that the AW of soils with different textures may be affected in different ways. Upon addition of 5% sludge, the AW of a sandy soil increased, whereas that of a loamy soil was unaffected (Metzger, 1986). This last finding lends support to the suggestion made by McRae and Mehuys (1985) that organic matter addition might increase AW only in very sandy soils, with low intrinsic AW.

Khaleel et al. (1981) proposed exponential multiple regression equations to describe the effect of the increase in OC content due to waste application, and of the sand percentage on the increase in water retention. Approximately 80% of the variations in water retention at field capacity and wilting point can be explained by these parameters. The above-reported proportional relation between the extent of increase in water retention and the amount of sludge added into soils is additional proof of the role played by the sludge organic particles themselves in holding water. In some instances, it was shown that in sludge-amended soils the extent of AW increase was positively correlated with the increase in OC content (Epstein, 1975; Metzger, 1986). An illustration of the AW-OC correlation is given in Figure 2, where the patterns of change with time of these two properties, for a sandy soil–sludge mixture (Figure 2A) and for a silt loam–sludge mixture (Figure 2B), are presented. The increase in AW content becomes insignificant after 75 days (sandy soil–sludge mixture) and 54 days (silt loam–sludge mixture), as a result of the rapid OC mineralization occurring during the initial stage of the incubation.

The enhancement of soil water retention capacity due to sludge application is application is explained at least partly by structural modifications, namely, increases in total porosity and the fraction of porosity involved in water storage (Guidi, 1981; Khaleel et al., 1981). The water retained inside the soil pores at wilting point is surface adsorbed, so the increased water retention at −15 bar MP is explained by an increase in surface area due to the organic matter addition (Gupta et al., 1977). The water retention increase in the low tension range (field capacity) may be partly attributed to an increase in the number of small pores (Khaleel et al., 1981).

B. Structure and Aggregation

1. Bulk Density

Sludge, like other organic materials, is less dense than the mineral fraction of soils. Thus, sludge application reduces the soil bulk density. Reduced bulk densities were measured in sludge-amended plots, regardless of the soil texture (Gupta et al., 1977; Kladivko and Nelson, 1979a; Chang et al., 1983). Three years after sludge incorporation, significant reductions in bulk density were

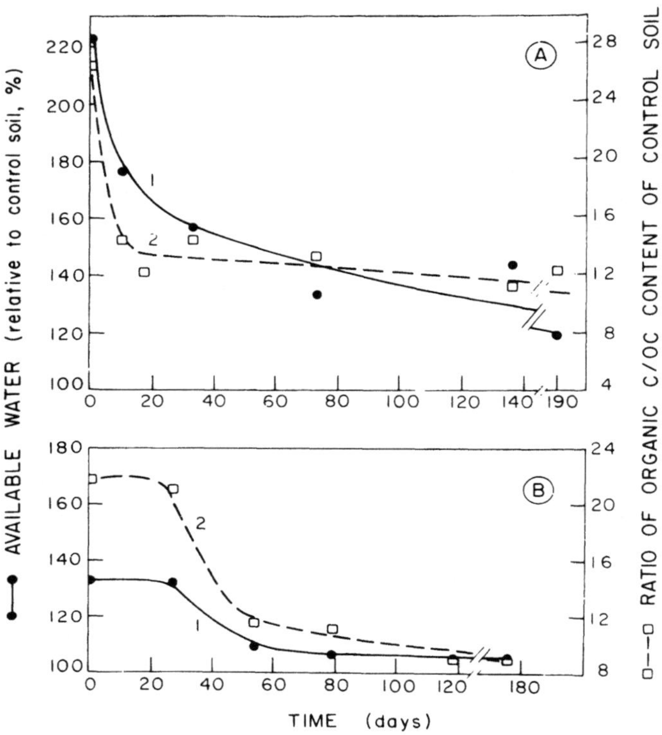

Figure 2. Changes occurring over 190 days of incubation in the available water (Curve 1) and the organic carbon content (Curve 2) in: (A) a sandy soil–undigested sludge (5%) mixture (from Metzger, 1986); and (B) a silt loam–digested sludge (5%) mixture (from Epstein, 1975).

found for three soils (clay loam, silt loam, and sandy loam), even at a low application rate of 27 t/ha (Hall and Coker, 1983). The bulk density of a sludge-amended sandy soil was linearly negatively correlated with the content of OC of sludge origin in the soil (Gupta et al., 1977). From the analysis of 12 studies dealing with the short- and long-term effects of application of organic wastes on bulk density, Khaleel et al. (1981) deduced a linear regression equation:

$$\Delta\rho b = 3.99 + 6.62\Delta \mathrm{OC} \qquad [3]$$

where $\Delta\rho b$ is relative reduction in bulk density (%) and ΔOC is relative OC increase (%).

Sludge OC may in addition affect indirectly the bulk density of soil, by improving the structural stability. Particularly in fine-textured soils, bulk density reduction may also be attributed to the increase in porosity resulting from better aggregation of the soil (Hall and Coker, 1983).

2. Porosity

A decrease in bulk density is automatically correlated with an increase in total porosity. Thus, sludge addition leads to an increase in the total pore volume of the soil (Gupta et al., 1977). Anaerobically and aerobically digested sludges, field-applied to a sandy loam soil, promoted a significant increase in the soil porosity. Six months after application of these sludges, the total porosity reached values ranging between 24% and 31% as compared with 17.5% for the control (Pagliai et al., 1981).

Besides total porosity, pore size distribution, which is more relevant with regard to the optimal physical status of soil, may be affected by sludge application. As a result of sludge and compost addition, the percentage of fissures (> 500 μm) decreased and the percentage of transmission (50–500 μm) and storage (0.5–50 μm) pores increased (Pagliai et al., 1981). According to Kladivko and Nelson (1979a), liquid sludge application to a slit loam promoted a decrease in bulk density associated mainly with an increase in the volume of "large" pores (> 50 μm). Wei et al. (1985) showed that 6 years after the application of dewatered sludge to a clay soil, "large-pore" space increased to the detriment of capillary pore space. On the whole, the above-reported results attest to an improvement in porosity following sludge application to soil.

3. Aggregation

Soil aggregation is defined mainly by the size-distribution of aggregates and their stability under wetting. In surface soils, organic matter is the main factor responsible for the water stability of soil aggregates. As aggregate stability is positively correlated with the soil OC content (Tisdall and Oades, 1982), it may be expected that the addition of materials rich in OC, such as sludge, will lead to an improvement of the aggregation status. Indeed, field as well as laboratory studies showed that the application of sludge to soil induces an increase in the number and size of water-stable aggregates (WSA). Figure 3

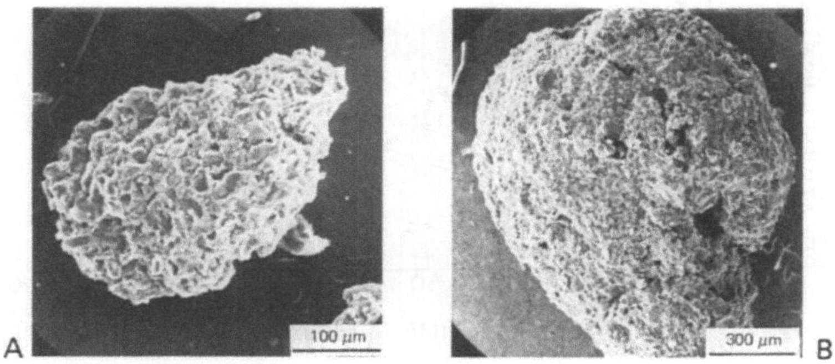

Figure 3. SEM micrographs representing the general aspect of water-stable aggregates isolated from: (A) a silt loam soil (control); and (B) a silt loam–sludge (5%) mixture after 16 days of incubation at 25°C.

illustrates the typical morphology of WSA (200–2,000 μm in diameter) formed in a loamy soil following sludge application, which were examined by scanning electron microscopy (SEM). As compared with control aggregates (Figure 3A), WSA found in the incubated sludge-amended soil (Figure 3B) are characterized by large size and a loose and porous internal structure. Incubation of soil–sludge mixtures at temperatures between 15°C and 35°C promoted a significant increase in the percentage of WSA (Epstein, 1975; Morel and Jacquin, 1977). Four months after moderate field applications of sludge, the mean-weight diameter of WSA was significantly increased (Kladivko and Nelson, 1979a). Other field experiments carried out on various soil types, under various climates, and with various types of sludge and compost, fully confirm these findings (Morel et al., 1978; Subbiah and Ramulu, 1979; Pagliai et al., 1981; Borchert, 1983). The aggregation pattern observed by us following addition of sludge to three different soils (Figure 4) exhibits three phases: a short 10-day period characterized by a rapid rate of WSA formation, then a period of the same duration displaying a moderate decrease in the WSA content, and finally a longer period characterized by a slow secondary increase in the WSA content. In this experiment, the enhancement in aggregate stability remained significant for at least 4 months for the sludge-amended soils. Under field conditions, the enhancement in structural stability in-

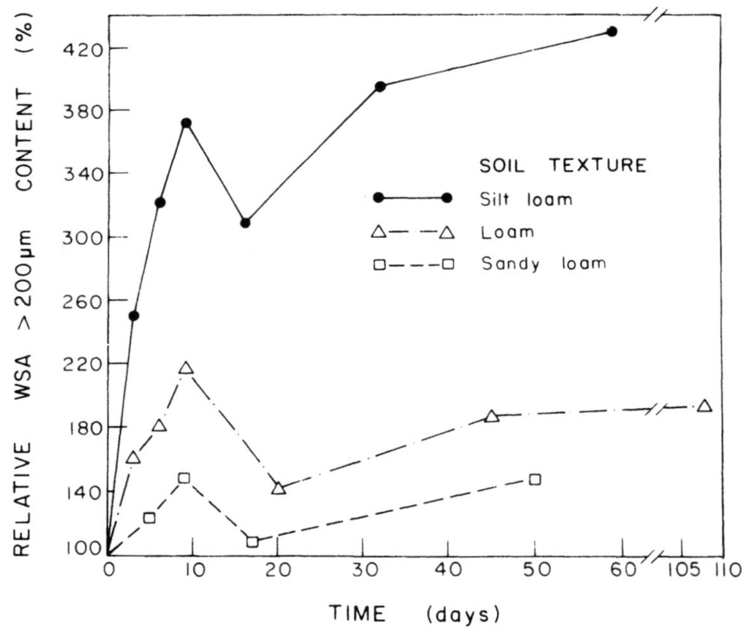

Figure 4. Dynamics of the aggregation status (expressed as the content of water-stable aggregates; WSA > 200 μm) in three types of soil after addition of 5% sludge (temperature of incubation: 25°C).

duced by sludge addition decreased with time as a result of the decay of the added OC (Kladivko and Nelson, 1979a). Long-lasting effects of sludge on structural stability were found, however, 4 years after the last date of a 140 t/ha application to a loamy clay (Morel and Guckert, 1983), and 6 years after a single 112 t/ha application to a silty clay loam (Wei et al., 1985).

The character of the increase in soil structural stability resulting from sludge application is similar to that induced by the addition to the soil of various simple or complex organic substances ranging from glucose to straw and peat (Morel and Jacquin, 1977). The microbial population developed on these organic substrates is responsible primarily for the formation and stabilization of aggregates. This viewpoint was confirmed by our experimental work, indicating that for soil–sludge mixtures in which microbial activity was suppressed, no aggregative processes were recorded (Metzger et al., 1986). Potential binding agents between soil mineral particles may include filamentous microbes (fungus hyphae), metabolites produced by the decomposers, as well as organic substances brought in with the applied waste (Guidi, 1981; Morel and Guckert, 1983). Our results show that both biological and chemical agents are at work in the aggregative processes found in soil–sludge mixtures. Observation under a scanning electron microscope of the WSA formed in a loamy soil following sludge application (Figure 5) showed the presence of three types of microorganisms (bacteria, Figure 5, C and F; actinomycetes, Figure 5D; and fungi, Figure 5E), as well as of cements of organic origin (Figure 5B). The relative contribution of the main groups of microorganisms in stabilizing aggregates remains a controversial issue. To clarify this point, the influence on the sludge-induced aggregative processes of bacteriocide and fungicide treatments was checked. The interpretation of the results obtained indicated that fungi make the dominant contribution to the increase in the WSA content following sludge application. Accordingly, the two main binding mechanisms that might be responsible for the aggregate formation and stabilization in sludge-amended soils are the cementation of primary particles by fungal polysaccharides and the physical entanglement of primary particles in the mesh of fungal hyphae (Metzger et al., 1986).

On the other hand, it was found that the water-soluble components of sludge are able to bind together well-defined soil constituents, like kaolinite and montmorillonite particles (Metzger and Robert, 1985). The organic hydrophobic fraction of sludge has also been proposed to be responsible for the aggregate stabilization. The fat and resin fraction of sludge was much more effective than the water-soluble fraction in increasing the water stability of soil aggregates (Guidi et al., 1983). A majority of studies regard polysaccharides, mostly by-products of the microbial activity developed on the C-containing substrate, as the main chemical aggregate-stabilizing agents. The positive correlation existing between structural stability and polysaccharide content in soils to which organic materials were added was extensively demonstrated (e.g., Guckert, 1973; Cheshire et al., 1984; Metzger et al., 1986). Figure 6

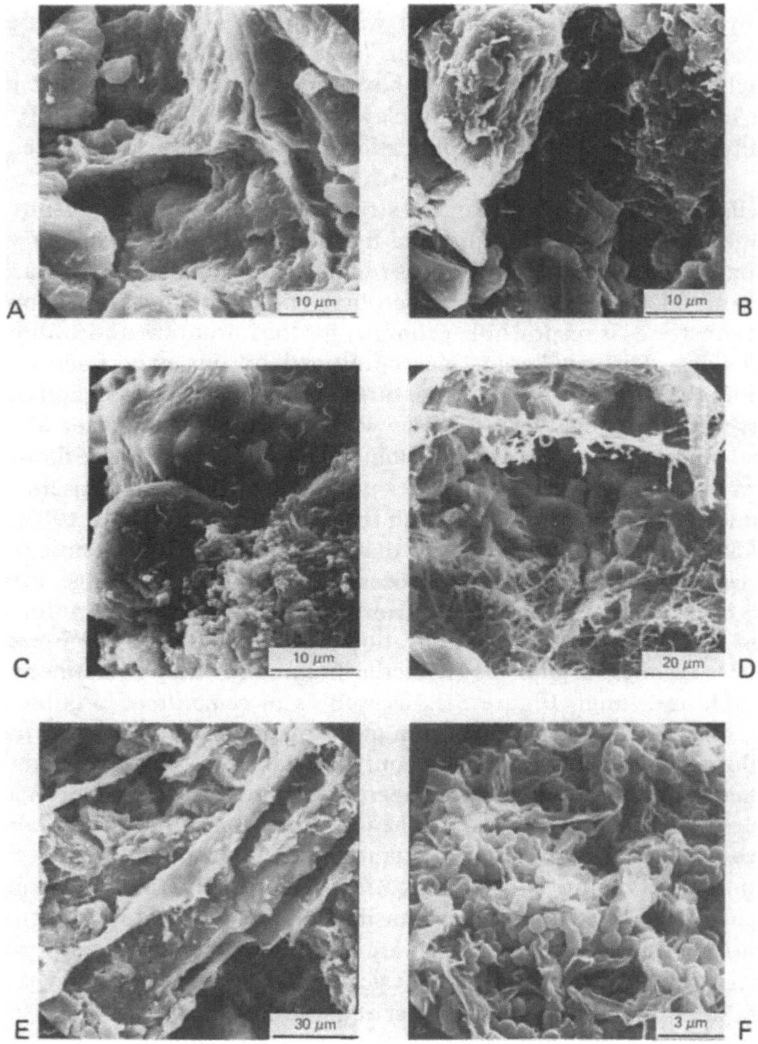

Figure 5. SEM micrographs representative of the internal structure of water-stable aggregates isolated from: (A) a silt loam soil (control); (B)–(F) a silt loam–sludge (5%) mixture after 16 days of incubation at 25°C.

illustrates the correlation existing in sludge-amended soils between the WSA content and the water-soluble polysaccharides content. It can be seen that for both soils studied, the patterns of change with time of these two properties are very similar. This lends support to the hypothesis that polysaccharides serve as the major cementing agents in aggregates, the stability of which is enhanced by adding sludge.

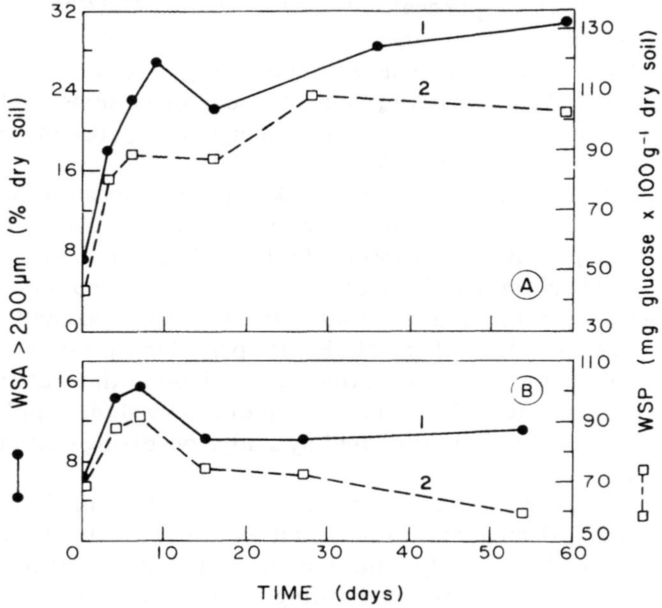

Figure 6. Changes over 60 days of incubation at 25°C of the water-stable aggregates (WSA, Curve 1) and water-soluble polysaccharides (WSP, Curve 2) contents in: (A) a silt loam–sludge (5%) mixture; and (B) a loamy soil–sludge (5%) mixture.

C. Water Transmission Properties

With a few exceptions, such as the temporary plugging of the uppermost soil layer as a result of liquid sludge application, the organic components of sludge are not considered to affect directly the water transmission properties of soil. Rather, they are thought to affect these properties indirectly, through their influence on the aggregation status, and hence on porosity. As good structural conditions are usually associated with adequate water transmission properties, it may be inferred from the previous paragraph that sludge will generally improve the water transmission properties.

1. Hydraulic Conductivity

The number of publications on the effect of sludge and other organic wastes on soil hydraulic conductivity (K) is limited and the reports provide questionable results because of the high variability inherent in the determination of K (Khaleel et al., 1981; Chang et al., 1983). However, most of the papers indicate increases in the saturated K (K_{sat}) after sludge application, usually explained by parallel decreases in the bulk density and increases in the total porosity (Khaleel et al., 1981). For soils with a wide range of textures, the K_{sat} in sludge-amended plots and sludge–soil mixtures was usually higher than the control (Epstein, 1975; Gupta et al., 1977; Morel et al., 1978; Chang et al., 1983; Wei et al., 1985). By adding sludge compost to a sandy soil, Jacobowitz and Steenhuis (1984) obtained increases in both K_{sat} and K_{unsat} (0 to -0.1 bar matric potential range). Large rates of sludge or compost additions are necessary to induce statistically significant changes in K_{sat} (Chang et al., 1983; Jacobowitz and Steenhuis, 1984; Wei et al., 1985).

Epstein (1975) and Metzger (1986) checked the modification of K_{sat} with time during comparable laboratory incubations of silt loam–sludge and sandy loam–sludge mixtures, respectively. In both cases, the K_{sat} increased shortly after the beginning of incubation to reach a maximum in the course of the 2nd week of incubation. Thereafter, the K_{sat} decreased and ultimately returned to the control level after a 50–80-day period in the former study, and after 25 days of incubation in the latter study. These authors attributed the temporary enhancement in K_{sat}, which was induced by sludge application, to the rise in aggregate stability promoted by sludge occurring during the same incubation period.

There are only a few reports on decreases in K_{sat} following the addition of sludges to soil. K_{sat} decreased during the 1st month of incubation when a silt loam was mixed with raw sludge and then incubated at 35°C (Epstein, 1975). A 50% decrease in K_{sat} was recorded during the 1st week of incubation when a loamy soil–undigested sludge was incubated at 35°C (Metzger, 1986). Application of raw solid or liquid sludge to soil columns subjected to leaching also led to decreases in K_{sat} (Subbiah and Ramulu, 1979). Sludge decomposition products, microbial metabolites or fungal mycelia, may be responsible for the plugging of the pores of these fine-textured soils, when amended with these sludges characterized by a relatively large content of easily degradable OC. Gupta et al. (1977) found that K_{unsat} was lowered as a result of the application of sludge to a coarse sand. According to these workers, the presence of hydrophobic water-repellent substances of sludge origin may explain the restriction of water flow under unsaturated conditions.

2. Infiltration

Infiltration through the soil surface depends on the hydraulic conductivity in the underlying soil. The application of organic wastes, such as sludge, may improve both the initial infiltration rate and the steady state infiltration rate (Khaleel et al., 1981). Surface-applied and disked-in liquid sludge increased the infiltration rate of sandy loam and silty loam soils checked 6 months after the application, but not enough to obtain significantly different results (Kelling et al., 1977; Kladivko and Nelson, 1979a). When cores of the sludge-amended soils were subjected to rain with the help of a rainfall simulator, lower surface runoff volumes were obtained. The maximal and most significant effect was observed for the soil with the finest texture (Kladivko and Nelson, 1979b). Simulation results indicated that 1–5% solid sludge amendment resulted in a 1.65–2.85-fold increase in cumulative infiltration of water in a mine soil (Younos and Smolen, 1983). However, the immediate effect of liquid sludge may be plugging of the soil and a reduction in the infiltration rate (Kelling et al., 1977). In a few cases, negative effects of sludge and other organic wastes upon infiltration rate and soil erosion were reported (Mazurak et al., 1975; Gupta et al., 1977; Khaleel et al., 1981). To summarize, erosion that results from transport of soil particles in cases of high runoff can be markedly reduced by sludge addition (Kladivko and Nelson, 1979b).

The beneficial effect of sludge on the water stability of soil aggregates explains why soils can be protected against crust formation and the resulting reduction in infiltrability by the application of sludge or other organic wastes. High-intensity rainfall brings about the formation of a crust in the case of soils with low structural stability, especially fine-textured soils poor in OC (Hall and Coker, 1983). A field experiment showed that crusting was reduced by application of sludge and pig slurry to sandy loam and silty loam soils. Scanning electron and light microscope studies carried out on thin sections of the soils showed that the organic wastes enhanced the macroporosity of the surface soils and thus helped to conserve their permeability (Pagliai et al., 1983).

D. Other Physical Properties

1. Mechanical Properties

Very little data are available about the effect of sludge on soil mechanical properties. Research on manuring led to reports on the importance of soil organic matter with regard to soil compaction. Heavy applications of OC-containing wastes have been shown to reduce soil compactibility. Accordingly, the work of Chang et al. (1983) on cores of soils to which sludge was applied showed that sludge addition led to a reduction in the modulus of rupture, which is indicative of a reduction in the cohesion and a lower sensitivity to compaction.

2. Thermal Properties

Organic matter has a lower thermal conductivity than the soil mineral constituents. Sludge addition will therefore bring about a decrease in soil thermal conductivity. Heat capacity on a dry weight basis is not modified by the addition of OC sources. To predict the effect of sludge on soil temperature in the field, the effects of the added OC on both the thermal characteristics and the water content of the soil must be considered. Any increase in the water content is associated with a lowering of the thermal conductivity and an increase in the specific heat on a wet-weight basis. As a consequence, sludge addition buffers the soil against sudden temperature fluctuations (Gupta et al., 1977). Jacobowitz and Steenhuis (1984) confirmed that maximal day temperatures are lower in a sludge-amended soil than in control soil.

E. Significance of Land Disposal of Sludge with Regard to Soil Physical Properties

On the basis of the research carried out until now, we can affirm that a consistent and, on the whole, positive change in soil physical properties results from sludge addition. However, it must be emphasized that in many reported works, it was difficult to measure any significant modifications of the soil physical properties when application rates were low and based primarily on agronomic considerations (Khaleel et al., 1981). To meet the N requirements of agronomic crops, it is recommended to apply sludge at a yearly rate between 7 and 25 t/ha, levels for which the organic matter effect may be negligible (USDA, 1978). Consequently, higher application rates raise problems of overfertilization and toxicity (Guidi, 1981). In spite of these serious limitations, the organic value of sludge in agriculture remains meaningful. Several field experiments showed that at least some of the soil physical properties are changed following even low, admissible applications of sludge (Kladivko and Nelson, 1979a; Pagliai et al., 1981). Otherwise, repeated low levels of sludge application may have long-term effects on the humus content and the physical status of the soil (Khaleel et al., 1981; de Haan, 1983). A one-time, high application rate is feasible when the objective of the sludge disposal is to restore nonproductive soils with very poor physical properties, like strip mine lands or dune soils (USDA, 1978).

IV. Conclusions

This review presents data that indicate that sewage sludge may be used as a soil conditioner, since risks of contamination of soils and waters by components of the sludge can be minimized by using appropriate application conditions. Both the amount and composition of the organic matter added to the soil affect the extent to which its physical properties are modified. Although sludge application may improve many of the major hydraulic and structural soil properties, the most significant effects were reported for the water retention capacity and the water stability of aggregates. The higher water retention in soils amended with sludge is due primarily to the high water-holding capacity of the sludge organic components. The formation and stabilization of aggregates induced by sludge can be explained by: (1) the direct interaction between sludge organic polymers and soil mineral components; (2) the stimulation of the microflora, resulting from the addition of this energetic substrate and the indirect interaction between microorganisms, microbial products, and soil components.

The beneficial influence of sludge decreases with time, at approximately the same rate as the sludge OC decomposes. Significantly increased soil humus content, and long-lasting improvement of the soil physical status can, however, be obtained by successive and moderate applications of sludge.

As the literature provides only scant information on the chemical and biological processes occurring in the soil–sludge system, further research work is needed to define more clearly these processes and the factors affecting them.

Acknowledgments

The financial support of the Israel National Research Council for Research and Development and the German Bundesministerium fur Forschung und Technologie (Research Grant WT 222) is gratefully acknowledged. Our thanks are extended to Professor C. Schlichting of the Höhenheim Universität Stuttgart FRG for his suggestions that were given to us during the performance of the experimental work.

References

Agbim, N.N., B.R. Sabey, and D.C. Markstrom. 1977. Land application of sewage sludge: V. Carbon dioxide production as influenced by sewage sludge and wood waste mixtures. *J. Environ. Qual.* 6:446–451.

Alexander, M. 1977. *Introduction to Soil Microbiology* (2nd ed.). Wiley, New York.

Borchert, H. 1983. Influence of increasing amounts of sewage sludge on the soil structure. In: G. Catroux, P. L'Hermite, and E. Suess (eds.), *The Influence of Sewage Sludge on Soil Physical and Biological Properties of Soils.* D. Reidel, Dordrecht, Holland, pp. 77–86.

Boyd, S.A., L.E. Sommers, and D.W. Nelson. 1980. Changes in the humic acid fraction of soil resulting from sludge application. *Soil Sci. Soc. Am. J.* 44:1179–1186.

Chang, A.C., A.L. Page, and J.E. Varneke. 1983. Soil conditioning effects of municipal sludge compost. *J. Environ. Eng. (ASCE)* 109:574–583.

Cheshire, M.V., C.P. Sparling, and C.M. Mundie. 1984. Influence of soil type, crop and air drying on residual carbohydrate content and aggregate stability after treatment with periodate and tetraborate. *Plant Soil* 76:339–347.

de Haan, S. 1983. General comments on the organic value of sludge. In: G. Catroux, P. L'Hermite, and E. Suess (eds.), *The Influence of Sewage Sludge Application on Physical and Biological Properties of Soil.* D. Reidel, Dordrecht, Holland, pp. 62–64.

Epstein, E. 1975. Effect of sewage sludge on some soil physical properties. *J. Environ. Qual.* 4:139–142.

Epstein, E., J.M. Taylor, and R.L. Chaney. 1976. Effects of sewage sludge and sludge compost applied to soil on some soil physical and chemical properties. *J. Environ. Qual.* 5:422–425.

Fine, P., U. Mingelgrin, and B. Yaron. 1986. Nitrogen mineralization and NH_3 volatilization in sludge-amended soil in relation to soil type and amendment ratio. In: *Soil Improvement through Land Disposal of Sludge* (Final Research Report to NCRD and BMFT). ARO, The Volcani Center, Bet-Dagan, Israel, pp. 121–148.

Gilmour, C.M., F.E. Broadbent, and S.M. Beck. 1977. Recycling of carbon and nitrogen through land disposal of various wastes. In: L.F. Elliott and F.J. Stevenson (eds.), *Soils for Management of Organic Wastes and Wastewater.* SSSA, ASA and CSSA, Madison, WI, pp. 172–194.

Gilmour, J.T., and C.M. Gilmour. 1980. A simulation model for sludge decomposition in soil. *J. Environ. Qual.* 9:194–199.

Guckert, A. 1973. *Contribution a l'Étude des Polysaccharides dans les Sols et Leur Rôle dans les Mécanismes d'Agrégation.* Doctoral thesis, Sci. Nat., University of Nancy, Nancy, France.

Guidi, G. 1981. Relationships between organic matter of sewage sludge and physico-chemical properties of soil. In: P. L'Hermite, and E. Ott (eds.), *Treatment and Use of Sewage Sludge.* D. Reidel, Dordrecht, Holland, pp. 530–544.

Guidi, G., G. Petruzelli, and M. Giachetti. 1983. Effect of three fractions extracted from sewage sludge on the water stability and surface area of soil aggregates. *Soil Sci.* 136:158–163.

Gupta, S.C., R.H. Dowdy, and W.E. Larson. 1977. Hydraulic and thermal properties of a sandy soil as influenced by incorporation of sewage sludge. *Soil. Sci. Soc. Am. J.* 41:601–605.

Hall, J.E., and E.G. Coker. 1983. Some effects of sewage sludge on soil physical conditions and plant growth. In: G. Catroux, P. L'Hermite, and E. Suess (eds.), *The Influence of Sewage Sludge Application on Physical and Biological Properties of Soil.* D. Reidel, Dordrecht, Holland, pp. 43–61.

Hiroux, G., G. Lefevre, and M.A. Gaffet. 1983. Composition organique et décomposition dans le sol de boues mixtes de station d'épuration urbaine. *Science du Sol (Bull. AFES)* 1:17–26.

Hohla, G.N., R.L. Jones, and T.D. Hinesly. 1978. The effect of anaerobically digested sewage sludge on organic fractions of Blount silt loam. *J. Environ. Qual.* 7:559–563.

Jacobowitz, L.A., and T.S. Steenhuis. 1984. Compost impact on soil moisture and temperature. *Biocycle* 25:56–60.

Kelling, K.A., A.E. Peterson, and L.M. Walsh. 1977. Effect of waste water sludge on soil moisture relationships and surface runoff. *J. Water Pollut. Control Fed.* 49:1698–1703.

Khaleel, R., K.R. Reddy, and M.R. Overcash. 1981. Changes in soil physical properties due to organic waste applications: A review. *J. Environ. Qual.* 10:133–141.

Kladivko, E.J., and D.W. Nelson. 1979a. Changes in soil properties from application of anaerobic sludge. *J. Water Pollut. Control Fed.* 51:325–332.

Kladivko, E.J., and D.W. Nelson. 1979b. Surface runoff from sludge-amended soils. *J. Water Pollut. Control Fed.* 51:100–110.

Lynch, J.M., and E. Bragg. 1985. Microorganisms and soil aggregate stability. In: B.A. Stewart (ed.), *Advances in Soil Science*, Vol. 2. Springer-Verlag, New York, pp. 133–171.

Mazurak, A.P., L. Chesnin, and A.E. Tiarks. 1975. Detachment of soil aggregates by simulated rainfall from heavily manured soils in eastern Nebraska. *Soil. Sci. Soc. Am. Proc.* 39:732–736.

McRae, R.J., and G.R. Mehuys. 1985. The effect of green manuring on the physical properties of temperate-area soils. In: B.A. Stewart (ed.), *Advances in Soil Science*, Vol. 3. Springer-Verlag, New York, pp. 71–94.

Metzger, L. 1986. *Effect of the Organic Components of Sewage Sludge on Soil Physical Properties.* Ph.D. thesis, The Hebrew University of Jerusalem, Israel.

Metzger, L., D. Levanon, and U. Mingelgrin. 1987. Effect of sludge on soil structural stability: Microbiological aspects. *Soil Sci. Soc. Am. J.* 51:346–351.

Metzger, L., and M. Robert. 1985. A scanning electron microscopy study of the interactions between sludge organic components and clay particles. *Geoderma* 35:159–167.

Miller, R.H. 1974. Factors affecting the decomposition of anaerobically digested sewage sludge in soil. *J. Environ. Qual.* 3:376–380.

Morel, J.L., and A. Guckert. 1983. Influence of limed sludge on soil organic matter and soil physical properties. In: G. Catroux, P. L'Hermite, and E. Suess (eds.), *The Influence of Sewage Sludge Application on Physical and Biological Properties of Soils.* D. Reidel, Dordrecht, Holland, pp. 25–42.

Morel, J.L., A. Guckert, and M. Sedogo. 1978. Effets de l'épandage des boues résiduaires urbaines sur l'état physique du sol. *Bull. ENSAIA (Nancy)* 20(1–2): 13–19.

Morel, J.L., and F. Jacquin. 1976. Etude le la minéralisation dans un sol agricole de boues résiduaries d'origines diverses. *Bull. ENSAIA (Nancy)* 23(1–2): 81–90.

Morel, J.L., and F. Jacquin. 1977. Détermination de l'aptitude ă la biodégradation de boues résiduaires d'origines diverses: Action sur les propriétés physicochimiques du sol. In: *Soil Organic Matter Studies*, Vol. II. IAEA, Vienna, pp. 265–275.

Morel, J.L., M. Sedogo, and F. Jacquin. 1977. Biodégradation et humification: IV. Influence de la technologie d'épuration sur l'evolution d'une boue résiduaire urbaine incorporée ă un sol neutre. *Science du Sol (Bull. AFES)* 3: 157–166.

Pagliai, M., E.B.A. Bisdom, and S. Ledin. 1983. Changes in surface structure (crusting) after application of sewage sludge and pig slurry to cultivated agricultural soils in Northern Italy. *Geoderma* 30: 35–53.

Pagliai, M., G. Guidi, M. La Marca, M. Giachetti, and G. Lucamante. 1981. Effects of sewage sludges and compost on soil porosity and aggregation. *J. Environ. Qual.* 10: 556–561.

Reddy, K.R., R. Khaleel, and M.R. Overcash. 1980. Carbon transformations in the land areas receiving organic wastes in relation to nonpoint source pollution: A conceptual model. *J. Environ. Qual.* 9: 434–442.

Schaumberg, G.D., C.S. Le Vesque-Madore, G. Sposito, and L.J. Lund. 1980. Infrared spectroscopic study of the water-soluble fraction of sludge-soil mixtures during incubation. *J. Environ. Qual.* 9: 297–303.

Sommers, L.E., D.W. Nelson, and D.J. Silviera. 1979. Transformation of carbon, nitrogen and metals in soils treated with waste materials. *J. Environ. Qual.* 8: 287–294.

Subbiah, S., and U.S.S. Ramulu. 1979. Influence of sewage wastes addition on the soil characteristics: I. Effect on physicochemical properties. *Mysore J. Agric. Sci.* 13: 408–414.

Terry, R.E., D.W. Nelson, and L.E. Sommers. 1979a. Carbon cycling during sewage sludge decomposition in soils. *Soil Sci. Soc. Am. J.* 43: 494–499.

Terry, R.E., D.W. Nelson, and L.E. Sommers. 1979b. Decomposition of anaerobically digested sewage sludge as affected by soil environmental conditions. *J. Environ. Qual.* 8: 342–347.

Tester, C.F., L.J. Sikora, J.M. Taylor, and J.F. Parr. 1977. Decomposition of sewage sludge compost in soil: I. C and N transformation. *J. Environ. Qual.* 6: 459–463.

Tisdall, J.M., and J.M. Oades. 1982. Organic matter and water-stable aggregates in soils. *J. Soil. Sci.* 33: 141–163.

USDA. 1978. *Improving Soils with Organic Wastes.* Report No. 1979/0-623-448/770. U.S. Government Printing Office, Washington, DC.

Wei, Q.F., B. Lowery, and A.E. Peterson. 1985. Effect of sludge application on physical properties of a silty clay loam soil. *J. Environ. Qual.* 14: 178–180.

Yaron, B., A.J. Vinten, P. Fine, L. Metzger, and U. Mingelgrin. 1984. The effect of solid organic components of sewage on some properties of the unsaturated zones. In: B. Yaron, G. Dagan, and J. Goldshmid (eds.), *Pollutants in Porous Media, Ecological Studies*, Vol. 47. Springer-Verlag, Berlin, pp. 162–181.

Younos, T.M., and M.D. Smolen. 1983. Simulating the hydraulic behavior of a sewage-sludge amended mine soil. *Trans. Am. Soc. Agric. Eng.* 26: 1397–1400.

Efficient Resource Management Systems for Drylands of India

J. Venkateswarlu*

* Central Research Institute for Dryland Agriculture, Hyderabad—500 659 (India).

Advances in Soil Science, Volume 7

(*cont.*)

I. Introduction

India is a vast country with different agro-climatic regions. About 70% of the 143.8 Mha of arable land depends entirely on natural precipitation for crop production. Much of this land is best described as dryland. It is important to distinguish between dryland agriculture in particular and rain-fed agriculture in general. Clearly, they both exclude irrigation. The emphasis for rain-fed agriculture, however, is often on disposal of excess water, maximum crop yields, high levels of fertilizer inputs, and water erosion constraints. In dryland agriculture, emphasis is on water conservation, sustainable crop yields, limited inputs of fertilizer, and both wind and water erosion constraints (Burnett, 1985). Drylands generally receive less than 750 mm of rainfall annually and are arid to semiarid. However, an absolute value cannot be stated because seasonal rainfall distribution and temperature affect crop water requirements and thus the kinds of soil and water management practices. Thus a region with annual rainfall of 750 mm with good monthly distribution in a cool climate might utilize rain-fed agriculture-type management systems, whereas in a hot climate a region with poorly distributed rainfall of less than 750 mm per year might appropriately require dryland farming techniques (Stewart and Burnett, 1985). Both dryland agriculture and rain-fed agriculture are considered in this review, but attention is focused on drylands.

In this review, the important soil groups in drylands of India are presented with a brief discussion on their characteristics in relation to crop production. Next, the existing systems of crop production and the enormous scope of increasing productivity are highlighted. This is followed by a brief discussion on evaluation of steps singly and jointly in crop production. The improved cropping systems are then discussed, relating them to the soils and rainfall, identifying the growing season, and suggesting mono-, inter-, and double-cropping systems for various agro-climatic regions. The effect of tillage and mulching is discussed with particular reference to soil problems and moisture conservation. A detailed discussion on fertilizer use in drylands is presented with greater emphasis on more efficient use of chemical fertilizers and the need for chain recuperation through organic recycling. This is followed by on-farm rainwater management for bringing more rainwater into the profile, coupled with a discussion on runoff collection and recycling. Finally, the need for stability in crop production in drylands is discussed and potential areas for different crops are presented. A brief discussion on agro-technology transfer is also presented.

II. Soils of Drylands in India

India has a large variety of soils and crops. Broadly speaking, the soils can be grouped as:

1. Sierozemic soils (Aridisols)
2. Submontane soils (Mollisols, Inceptisols, and Entisols)
3. Alluviums (Inceptisols, Alfisols, and Entisols)
4. Red soils (Alfisols)
5. Black soils (Vertisols)

Important soil characteristics in relation to crop production are discussed in the following paragraphs. The profile analysis and nomenclature were obtained from Murthy et al. (1981) and Soil Survey staff (1978).

A. Sierozems (Aridisols)

Camborthids and Solorthids are the major groups in Aridisols. These soils are extremely light-sandy, loamy sand, or sandy loams. Calcium carbonate occurs at various depths influencing the effective depth of the soil. Crusting is a serious problem in these soils. Soil erosion by wind is even more serious. The thickness of the solum varies from 70 to 100 cm.

Because the soils are very sandy, they hold only low amounts of water and nutrients. They are universally poor in nitrogen and sometimes inadequate even in phosphorus. Subsoil salinity is not uncommon due to extreme aridity. The predominant clay minerals are smectite, mica, and kaolinite.

The texture largely determines the soil water storage capacity. In deep sandy loams, either a rainy or postrainy crop can be produced. But in loamy sands or sand, only rainy season cropping is possible.

The major characteristics of typical Camborthid profile from Jodhpur are presented in Table 1. These soils (Chirai series) were developed on old alluvium modified by aeolian material and occur in Jodhpur and Bikaner districts of Rajasthan. Usually the cambic horizon is encountered below 10–25 cm and extends upward to the zone of calcium accumulation. The above profile is from a Typic Camborthid of Jodhpur.

B. Submontane Soils (Mollisols, Inceptisols, and Entisols)

Hapludolls (Mollisols), Haplaquepts (Inceptisols), and Udorthents (Entisols) are more common in the region occurring on level to gently sloping lands, lower elements of slopes, and steep slopes, respectively. These soils are generally deep, the texture being silty loam to loam, and hold as much as 300 mm of water per meter of profile. Land slides and soil erosion are common.

Table 1. Major characteristics of Camborthids

Soil horizon	Depth (cm)	Clay (%)	pH (1:2.5)	EC ($dS \cdot m^{-1}$)	$CaCO_3$ (%)	Organic carbon (%)	Exchangeable bases ($mmol \cdot kg^{-1}$)	CEC ($mmol \cdot kg^{-1}$)
Ap	0–9	9.2	8.0	0.40	1	0.11	59	61
A_{12}	9–36	11.2	8.1	0.41	8	0.09	62	65
B	36–76	12.4	8.3	0.37	6	0.10	51	59
B_2Ca	76–90	9.6	8.3	0.40	9	0.06	50	54
C_1Ca	90–140	12.7	8.4	0.55	10	0.06	53	54
C_2Ca	140–170	13.4	8.4	0.50	12	0.06	50	59

Source: Murthy et al. (1981).
EC = Electrical; CEC = Cation exchange capacity.

Table 2. Major characteristics of Hapludolls

Horizon	Depth (cm)	Clay (%)	pH (1:2.5)	$CaCO_3$ (%)	Organic carbon (%)	Exchangeable bases ($mmol \cdot kg^{-1}$)	CEC ($mmol \cdot kg^{-1}$)
Ap	0–20	17.8	7.1	—	0.71	63	74
A_{12}	20–27	17.5	7.1	—	0.72	70	80
A_3	27–43	17.8	7.0	—	0.48	74	80
B_{21}	43–67	18.7	7.0	—	0.42	80	90
B_{22}	67–80	16.0	7.1	—	0.33	76	89
B_3	80–106	17.5	7.3	—	0.20	74	91
C_1	106–119	11.7	7.5	Tr	0.21	57	59
IIC_2	119–128	11.5	7.7	Tr	0.15	52	55

Source: Murthy et al. (1981).
EC = Electrical conductivity; CEC = Cation exchange capacity.

They respond well to nitrogen. Phosphorus becomes limiting only at high production levels. In lower reaches (Haplaquepts), water stagnation is possible, which calls for special efforts in nutrient management, particularly of nitrogen. The predominant clay minerals are illite, smectite, and chlorite/smectite.

The analysis of a typical Hapludoll is given in Table 2. These soils (Haldi series) occur on gently sloping lands in the Terai region of Uttar Pradesh. They are deep. The surface and subsurface layers have dark columns with low chromas and more than 1.0% organic matter. The above profile is from Nainital, Uttar Pradesh, and is Typic Hapludoll.

C. Alluviums (Inceptisols, Alfisols, and Entisols)

The more common groups are Ustocherpts (Inceptisols), Ochraqualfs (Alfisols), and Ustifuluvents (Entisols). These soils are generally loamy sands or sandy loams, depth not being a constraint. With low clay content, they hold less water (0.1–0.2 kg of water per kg of dry soil). The nutrient-holding capacity is also less. They are neutral to alkaline in reaction. Water intake is high. Crusting could be serious due to either a beating rain or high temperatures. This might accelerate the problem of runoff with subsequent rains. The slopes are mostly incipient.

The productivity of these soils is high, but they require liberal applications of nitrogen for high yields. Phosphorus is needed only for very high yields. Potassium is adequate. The predominant clay minerals are illite, smectite, kaolinite, and chlorite.

The lower reaches of the catena develop alkalinity and salinity. As the soils become heavier in texture, the water-holding capacity increases, leading to the scope for taking two crops in a year.

The profile analyses of Ustochrepts and Ochraqualfs are given in Tables 3 and 4. Ustochrepts soils (Table 3) occur on gently sloping alluvial terraces in Punjab, Haryana, and Delhi. They are deep. These soils are located on

Table 3. Major characteristics of Ustochrepts

Horizon	Depth (cm)	Clay (%)	pH (1:2.5)	EC ($dS \cdot m^{-1}$)	$CaCO_3$ (%)	Organic carbon (%)	Exchangeable bases ($mmol \cdot kg^{-1}$)	CEC ($mmol \cdot kg^{-1}$)
Ap	0–19	15.9	8.1	0.53	Tr	0.40	91	98
A_3	19–31	14.1	8.1	0.44	Tr	0.28	88	89
B_{21}	31–45	30.5	7.9	0.37	Tr	0.30	142	143
B_{22}	45–73	29.9	7.7	0.39	Tr	0.25	136	143
B_{23}	73–105	32.9	7.6	0.43	Tr	0.25	116	117
B_3	105–136	28.1	7.7	0.47	Tr	0.16	128	130
C	136–156	20.1	7.7	0.38	Tr	0.10	118	117

Source: Murthy et al. (1981).
EC = Electrical conductivity; CEC = Cation exchange capacity.

Table 4. Major characteristics of Ochraqualfs

Horizon	Depth (cm)	Clay (%)	pH (1:2.5)	EC ($dS \cdot m^{-1}$)	Organic carbon (%)	Exchangeable bases ($mmol \cdot kg^{-1}$)	CEC ($mmol \cdot kg^{-1}$)
Ap	0–8	22.5	6.6	0.33	0.42	101	116
A_3	8–30	30.4	6.6	0.37	0.21	127	137
B_{21}	30–81	38.9	6.7	0.30	0.21	147	151
B_{31}	81–94	39.0	7.7	0.64	0.24	153	157
II Ca	94–132	24.8	7.9	0.54	0.19	91	93

Source: Murthy et al. (1981).
EC = Electrical conductivity; CEC = Cation exchange capacity.

mid- and upper terraces. The development of a cambic horizon is prominent below 15–25 cm. There are evidences of moderate structural development and illuviation. The profile described is from Patiala (Punjab) and is a Udic Ustochrept (Nabha series).

Ochraqualf soils as characterized in Table 4 are found on Gangetic alluviums and occur on flat depressions in Varanasi, Gazipur, and Ballia districts of Uttar Pradesh. They are imperfectly drained as they occur in the lower elements of the topography and possess hydromorphic conditions. The soil color will be usually brownish grey with lower chromas. Evidence of agrillans is prominent. The described profile is an Acric Ochraqualf.

D. Red Soils (Alfisols)

The most prominent groups are Haplustalfs, Rhodustalfs, and Paleustalfs. They are usually encountered on mild slopes of interfluves and higher terraces.

These are more weathered soils. They are generally low in clay content and the predominent clays are kaolinite and illite. Most of them are sandy loams with an illuvial of higher clay below the plow layer. Due to illuviation, the subsoil is more compact, interspersed with saprolite. These soils tend to crust after a beating rain or with high temperatures. Crusting leads to higher runoff.

These soils are universally deficient in nitrogen and often in phosphorus as well. Potassium may be limiting with high production systems. Sometimes, calcium can be deficient, more so when the soils are monocropped with high calcium-demand crops like peanuts (*Arachis hypogaea* L.)

The Alfisols developed on granites are generally moderately deep (e.g., Anantapur in Andhra Pradesh), while the depth tends to increase with those developed on chernokites, limestone, quartzite, and sandstone. Incidentally, it has been observed that as the rainfall increases, the solum thickness also tends to increase. The profile water storage is about 0.15 $m^3\ m^{-3}$. As the clay increases, it might move up to 0.20 $m^3\ m^{-3}$.

Analysis of two typical profiles are given in Tables 5 and 6. The Pelatherai Haplustalfs (Table 5) were developed on calcic gneiss and occur on upper and middle pediments in Coimbatore, Salem, and Dhrmapuri districts of Tamil

Table 5. Major characteristics of Haplustalfs

Horizon	Depth (cm)	Clay (%)	pH (1:2.5)	Organic carbon (%)	$CaCO_3$ (%)	Exchangeable bases (mmol·kg^{-1})	CEC (mmol·kg^{-1})
Ap	0–12	13.5	8.4	0.66	2.1	129	130
B_{21}	12–39	20.9	8.3	0.74	1.6	183	236
B_{22}	39–57	22.9	8.6	0.81	2.1	212	256
C1 Ca	57–68	21.5	8.8	0.73	12.4	193	215

Source: Murthy et al. (1981).

EC = Electrical conductivity; CEC = Cation exchange capacity.

Table 6. Major characteristics of Paleustalfs

Horizon	Depth (cm)	Clay (%)	pH (1:2.5)	Organic carbon (%)	Exchangeable bases	CEC
					(mmol · kg^{-1})	
A_1	0–9	19.9	5.1	0.26	47	85
A_3	9–30	26.6	5.5	0.23	62	88
B_1	30–48	31.8	6.3	0.13	69	91
B_{21t}	48–73	22.5	6.3	0.10	86	96
B_{22t}	73–91	33.7	6.5	0.10	89	104
B_{23t}	91–114	33.9	6.5	0.08	94	114
B_{24t}	114–141	37.6	6.5	0.07	106	133
B_{3t}	141–186	40.6	6.3	0.05	118	143

Source: Murthy et al. (1981).
EC = Electrical conductivity; CEC = Cation exchange capacity.

Nadu. They are light in texture at the surface and the clay content increases with the depth, reaching the maximum in the agrillic horizon. The base saturation is high. The described profile is a Typic Haplustalf.

Soils of the Pusaro series (Table 6) were developed on weathered granite-gneiss and occur on peneplained plateaus in the Santal Parganas district of Bihar. They have deep profiles with prominent evidences of illuviated clay. The clay content is high throughout the 2-m profile. The profile described is a Ustic Paleustalf.

E. Black Soils (Vertisols)

The important groups are Chromusterts and Pellusterts. Vertisols are more productive than Alfisols. As the soil organic matter is low in Indian Vertisols, the productivity is largely dependent on the quantity of clay. These soils vary both in depth (15–240 cm) and clay content (30–70%). They are rich in bases. The predominent clay minerals are smectite, montmorillonite, nontronite, and beidellite. They hold more water (0.30–0.35 $m^3 m^{-3}$) than other predominent soils and are alkaline in reaction. With semiarid conditions, subsoil salinity is common. Boron toxicity can also be a problem.

Due to the high clay content, these soils tend to crack on drying. In low-rainfall areas (e.g., Bellary in Deccan), cracking occurs in late December when the conventional long-duration crops are at flowering stage. This leads to increased loss of soil moisture. Consequently, the long-duration crops suffer.

These soils have a tendency for sheet erosion. The runoff varies from 10% to 40%, increasing with increase in rainfall. Many deep soils are calcareous. Heavy structure and water logging make it difficult for early management of these soils. To cope with hard pans and/or increased weed infestation, these soils are commonly deep plowed (>25 cm) once in 3 years.

Profile analyses of typical Chromosterts and Pellusterts of subhumid and

Table 7. Major characteristics of Chromusterts

Horizon	Depth (cm)	Clay (%)	pH (1:2.5)	EC ($ds \cdot m^{-1}$)	Organic carbon (%)	$CaCO_3$ (%)	Exchangeable bases ($mmol \cdot kg^{-1}$)	ESP
Ap	0–11	56.1	8.0	0.1	0.40	0.9	532	7
A_{12}	11–29	57.9	8.1	0.1	0.34	2.7	449	9
A_{13}	29–54	58.1	8.1	0.1	0.34	3.3	572	7
A_{14}	54–95	59.0	8.1	0.1	0.30	2.3	581	8
A_{15}	95–119	55.5	8.1	0.1	0.30	1.9	452	10
A_{16}	119–147	52.0	8.2	0.1	0.27	3.3	492	10
A_c	147–160	52.5	8.2	0.1	0.27	14.9	432	11

Source: Murthy et al. (1981).

ESP = Exchangeable sodium percentage; EC = Electrical conductivity; CEC = Cation exchange capacity.

Table 8. Major characteristics of Pellusterts

Horizon	Depth (cm)	Clay (%)	pH (1:2.5)	Organic carbon (%)	$CaCO_3$ (%)	Exchangeable bases ($mmol \cdot kg^{-1}$)	ESP
Ap	0–12	64.6	8.0	0.97	3.4	658	2.6
A_{12}	12–60	64.7	8.0	0.70	3.3	665	3.7
A_{13}	60–112	68.2	8.0	0.65	3.4	672	5.7
A_{14}	112–150	69.3	8.0	0.53	3.4	668	5.1

Source: Murthy et al. (1981).

ESP = Exchangeable sodium percentage; EC = Electrical conductivity; CEC = Cation exchange capacity.

semiarid regions, respectively, are given in Tables 7 and 8. Soils of the Sarol series, described in Table 7, were developed in basaltic alluvium and occur on piedmont plains of the Malwa Plateau. They predominently exhibit Vertic characters and have high bases dominant in calcium and magnesium. The profile is a Typic Chromostert.

Soils of the Raichur series, described in Table 8, are found in old alluviums and occur generally on gently sloping lands in the Raichur district of Karnataka. They are mostly derived from granite and granatoides. Vertic characters are prominent and the chroma is low compared to the Chromostert associates. The profile is a Typic Pellustert.

III. Untapped Yield Potential in Drylands of India

The low yields in the existing systems of farming in drylands of India are primarily due to:

1. Use of conventional varieties of crops that have low yield potential
2. Less adequate attention on timely operations like sowing and weeding

3. Lack of response to fertilizer by the local varieties, and
4. Improper land management systems

The local varieties are long in duration and undergo any stress and strain that occurs in the rainy season due to improper distribution of rainfall. In the postrainy season, the varieties are too long to mature on the stored soil moisture, particularly in the Vertisols of the Deccan. The local varieties are shallow rooted and are surface feeders. They are also bred to perform as individual plants, which means they perform well only at low plant population per unit area.

Earlier research efforts were made during the 1930s with the intention of improving crop yields in drylands. Kanitkar et al. (1968) summarized the results of these efforts termed as Madras/Deccan/Bombay dry farming methods. They largely included:

1. Bunding with bund former to conserve soil and water
2. Deep plowing once in 3 years for better intake and storage of rainfall
3. Use of feedyard manure (FYM) to supply plant nutrients
4. Use of low seed rate
5. Intercultivation of crops for efficient utilization of the limited soil water

These practices gave a 15–20% increase in the yield over a base yield of 200 to 400 kg ha^{-1}. In other words, the actual yield increase varied from 30 to 80 kg ha^{-1}. Although the percentage yield increase was significant, it was not highly visible. So the farmers were not enthused with the results.

By the mid 1950s, the emphasis shifted to soil management. Soil and water conservation programs focused on contour bunding. But such an effort, sometimes, gave negative results, particularly so in Vertisols. Water congestion and washout were the problems. Even when yield increases were observed, they were again not more than 15–20% over the existing yields of 200–400 kg ha^{-1}. As before, there was no enthusiasm by the farmers in accepting these results.

The research continued and led to various other measures like graded bunds, bench terraces, graded border strips, and so forth. To contain erosion and increase groundwater recharge, gully plugging, check dams, percolation tanks, and gabion structures were developed.

At the same time the need for shorter duration crops was identified for a better fit into the dryland environments. And it was during the mid-1960s that the answer was available through hybrids and high-yielding varieties of crops, particularly coarse cereals. These high-yielding hybrids and varieties were not only responsive to fertilizer but also to management. There were great hopes and rekindled interest.

Through the establishment of the All India Co-ordinated Research Project for Dryland Agriculture (AICRPDA) in 1970, considerable data were generated on the untapped yield potential of these lands (Table 9). Similar results became available from the International Crops Research Institute for the Semi-Arid Tropics (ICRISAT), started in 1972.

Table 9. Yield of cereals in different soil types

Soil type	Center	Crop	Yield (Mg ha^{-1}) Control (farmers' practices)	Improved practices
Aridisols (sierozems)	Jodhpur	Pearl millet	0.3	1.8
(arid)	Hissar	Pearl millet	0.3	3.4
Inceptisols (submontane)	Samba	Corn	1.0	1.0
(subhumid)	Dehra Dun	Rice	1.2	4.4
Entisols, Inceptisols (Alluviums)				
(subhumid)	Varanasi	Rice	0.7	2.8
(semiarid)	Agra	Pearl millet	0.7	2.0
Vertisols (subhumid)	Indore	Corn	1.2	4.3
(semiarid)	Akola	Sorghum	0.5	1.8
	Kovilpatti	Sorghum	0.3	2.4
	Solapur	Sorghum	0.3	2.1
	Bellary	Sorghum	0.3	2.6
Alfisols (subhumid)	Ranchi	Rice	1.2	4.4
	Bhubaneswar	Rice	1.1	2.1
(semiarid)	Hyderabad	Sorghum	0.4	3.6
	Anantapur	Sorghum	0.3	1.4
	Bangalore	Finger millet	1.0	3.7

Source: Venkateswarlu (1981).

The data clearly indicated that with the existing practices the yield levels were very low varying from 0.3 Mg ha^{-1} in resource-poor situations to 1.0 Mg ha^{-1} under better soil and rainfall conditions. But with the improved practices, there was greater hope of increasing the productivity even to eight times the existing yields, particularly in optimal resource situations.

The improved practices included:

1. Proper tillage
2. Early sowing
3. Choice of suitable crops and varieties
4. Optimum plant population
5. Timely weeding and pest control
6. Moderate use of fertilizers

The Indian farmers most often have poor resources, and particularly so in the drylands. More than 70% of the farming community is of this nature. A crop production technology cannot be a package that costs two or three times more than the traditional system of farming. So a tier-system of crop production is needed. In other words, the contribution of individual components in crop production must be evaluated. Some data on these aspects are given in Table 10. The data clearly show that:

Table 10. Contribution of individual components and their combinations in dryland crop production

			Percent increase due to improved				
Center	Crop	Yield (Mg ha^{-1}) of conventional system	Seed	Management	Fertilizer	Seed + fertilizer	Seed + fertilizer + management
Anantapur (Alfisols)	Foxtail millet	0.36	38.9	13.9	36.7	88.8	100.0
Hyderabad (Alfisols)	Sorghum	0.90	91.1	7.8	92.2	211.1	386.7
	Castor	0.34	52.9	32.3	35.5	158.8	232.3
Solapur (Vertisols)	Sorghum	0.85	43.5	31.8	72.9	111.8	217.6
Rajkot (Vertisols)	Pearl millet	1.34	30.6	20.9	32.1	31.3	27.6
Kovilpatti (Vertisols)	Pearl millet	0.48	62.5	—	62.5	52.1	120.8
Bhubaneswar (Alfisols)	Rice	1.18	10.5	8.5	13.6	63.6	172.5
ICRISAT (Alfisols)	Sorghum	0.85	34.5	36.8	33.4	63.6	172.5
ICRISAT (Vertisols)	Sorghum	1.51	52.1	25.1	38.4	126.7	177.8
Average			46.3	22.1	46.3	96.4	164.5

Source: Annual Reports of respective centers and ICRISAT.

1. Efficient management, which is possible at the farm level, *per se* contributed up to 22% increase in yield of crops.
2. Seed and fertilizer each contribute as much as 40% and 50% increase in yield, respectively.
3. Even more important is the synergism when seed and fertilizer plus management lead to 164% increase in yields.

Simply stated, there is no substitute for efficient management in dryland crop production. And this should form the basic first step in improved crop production in drylands. This can be followed by improved seed and moderate levels of fertilizers.

IV. Cropping Systems

The drylands are usually monocropped. That is, the cropping intensity is 100. But with a careful analysis of the environment (soils and rainfall), it has now become possible to suggest different cropping intensities in different locations.

A. Soils

The soils hold water and nutrients for use by plants. The conventional available water capacity (AWC) is defined as the difference in volumetric water content between −0.33 bar and −15 bar of soil matric potential. A rooting depth may be assumed or measured to provide an estimate of total AWC in the profile.

But even below the −15-bar matric potential, crops are known to deplete soil water. Prihar et al. (1978) found that wheat (*Triticum aestivum* L.) was able to deplete soil water well beyond −15 bar, even to a depth of 2 m, in an Entisol. Similarly, chickpea (*Cicer arietinum* L.) removed substantial amounts of water from as deep as 150–180 cm and depleted soil water to below −15-bar value in an Entisol (Sandhu et al., 1978). In a Vertisol of the Deccan (Bellary), crops like *Dolichos lab lab* survived even at 0.04 kg per kg of soil moisture level, while the Permanent Wilting Coefficient (PWC) of these soils is 0.22–0.28 kg per kg. Thus the plant available water capacity (PAWC) becomes important. It was defined by Gardner et al. (1983) as the difference between field-measured upper storage limit and the lower storage limit measured in the field and summed over the plant-rooting depth. PAWC may vary with crop, stage of growth, level of plant stress, and so forth. (Smith et al., 1984).

Taking all of these views into consideration, the upper storage limit becomes important, and the amount of water that is so held depends on the depth and texture. Some important soil types and the maximum storage limits are given in Table 11.

Table 11. Important soil types, depth, water-holding capacity, and texture in drylands of India

Region	Depth	Upper limit in moisture storage (0.33 bar) (kg water kg^{-1} dry soil)	Texture
Inceptisols (Dehra Dun)	120	0.25–0.27	Silty loam to silty clay
Aridisols (Jodhpur)	200	0.07–0.14	Loamy sand
Entisols (Varanasi)	130	0.24–0.35	Silty loam–sandy loam
Vertisols (Raichur)	150	0.32–0.35	Clay to clay loam
Alfisols (Patancheru)	90	0.11–0.15	Sandy loam

Source: Annual Reports (1971–1972) of Dehra Dun and Jodhpur centers, and Murthy et al. (1981).

The available water properties for four soils (Figure 1) were determined by Russell (1978). In the deep Vertisol, available soil water was 230 mm in a 187-cm-deep profile, while in a medium Vertisol of 157 cm, it was 160 mm. In a deep Alfisol, it was 140 mm, compared to 95 mm for a medium Alfisol. Thus moisture storage values depended largely on the texture and depth of soils (Figure 2), as also depicted by Virmani et al. (1978).

B. Rainfall

Rains constitute the only source of water to dryland crops. So the total rains, their distribution, and the potential evapotranspiration are important to determine the period of dependable water availability in a given region. Hargreaves (1975) has proposed that the water requirement of dryland crops is met when the ratio of rainfall to potential evapotranspiration is at least 0.33. The ratio is also called moisture availability index (MAI) when the dependable amount of rainfall is considered as the criterion. In this context, Hargreaves (1975) considered that a dependable rainfall is one where the rainfall occurs with a probability of more than 75%.

Based on the ratio of actual evapotranspiration (AE) and the potential evapotranspiration (PE), Singh (1978) considered the weeks that had a value of AE:PE above 0.5 as dependable and identified a continuous number of these weeks. This period was called "the growing season" (Table 12). In fact, this is based on the proposals of Cocheme and Franquin (1967), who used such indices to describe the weather in a given region.

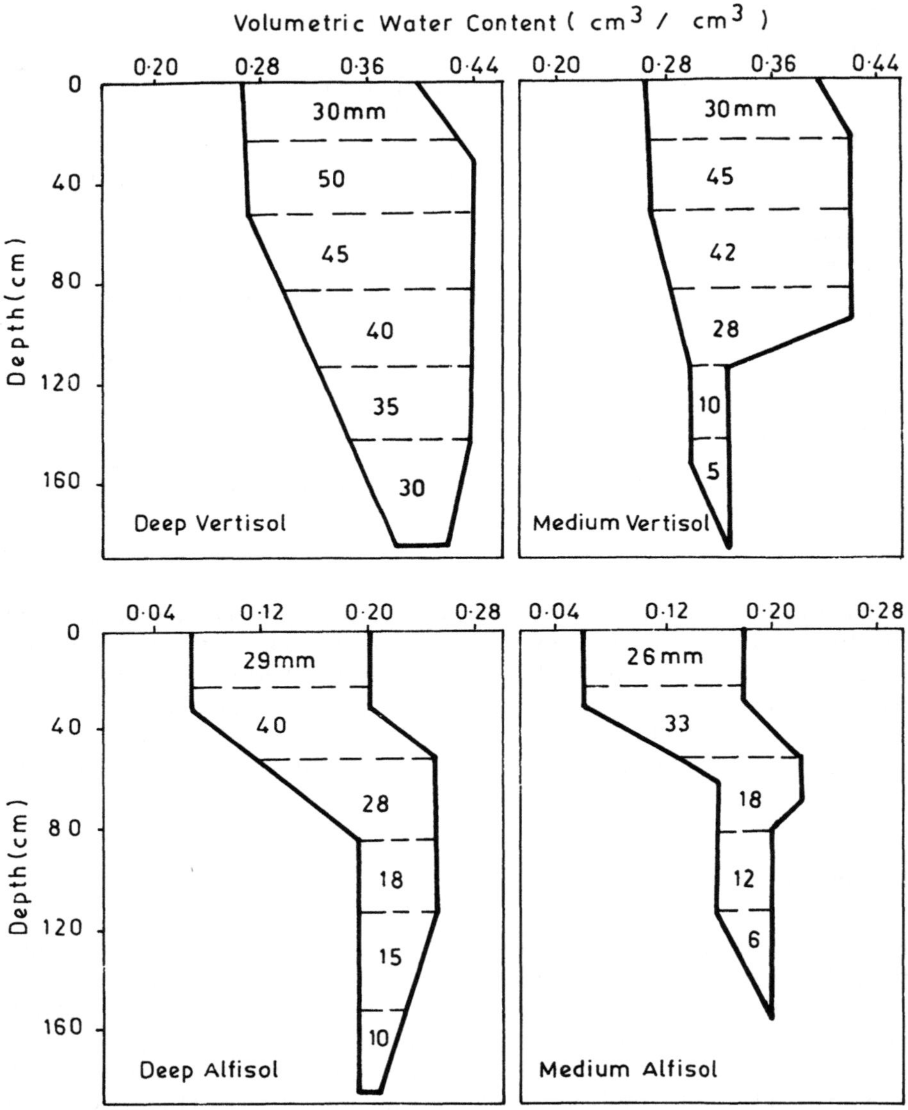

Figure 1. Available water profiles for four soils. From Russell (1978).

C. Water Balance Approach

For planning proper agricultural strategies, a combination of the rainfall data with the water-holding capacities of soils would be useful. They help in identifying the growing season with assured moisture availability. In Hyderabad with 760 mm of rainfall, Virmani et al. (1978) pointed out that the length of the growing season could vary from 15 to 23 weeks, depending on the soil

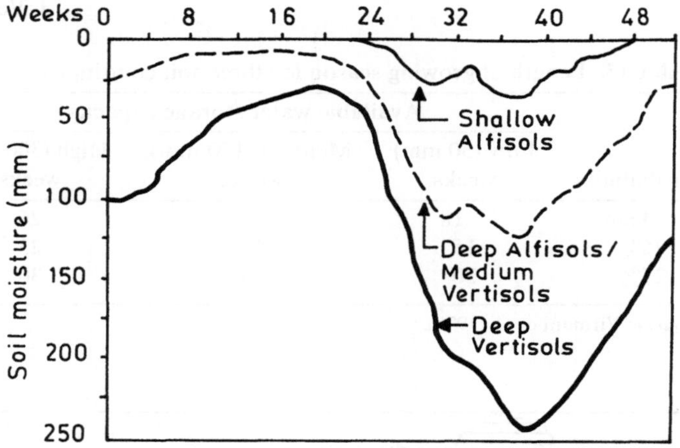

Figure 2. Weekly soil moisture storage in three soils (Hyderabad, 1901–1970). From Virmani et al. (1978).

Table 12. Length of effective cropping season at research centers of Dryland Project

Soil type and center	Standard weeks of assured moisture supply	Total duration in weeks
Aridisols (Sierozems) (semiarid)		
Jodhpur	27–37	11
Hissar	27–39	13
Mollisols, Inceptisols, Entisols (submontane soils) (subhumid)		
Samba	26–17	44
Dehra Dun	23–21	51
Inceptisols, Alfisols, Entisols (Alluviums)		
Varanasi (subhumid)	26–5	32
Agra (semiarid)	27–50	24
Vertisols		
Subhumid		
Indore	24–7	36
Semiarid		
Akola	24–50	27
Kovilpatti	38–6	21
Solapur	23–45	23
Bellary	37–44	8
Alfisols		
Subhumid		
Ranchi	22–52	31
Bhubaneswar	22–12	45
Semiarid		
Hyderabad	25–46	22
Anantapur	33–45	13
Bangalore	20–51	32

Source: Singh (1978).

Table 13. Length of growing season for three soil conditions

	Available water storage capacity		
Probability	Low (50 mm) weeks	Medium (150 mm) weeks	High (300 mm) weeks
Mean	18	22	26
75%	15	19	23
25%	20	24	30

Source: Virmani et al. (1978).

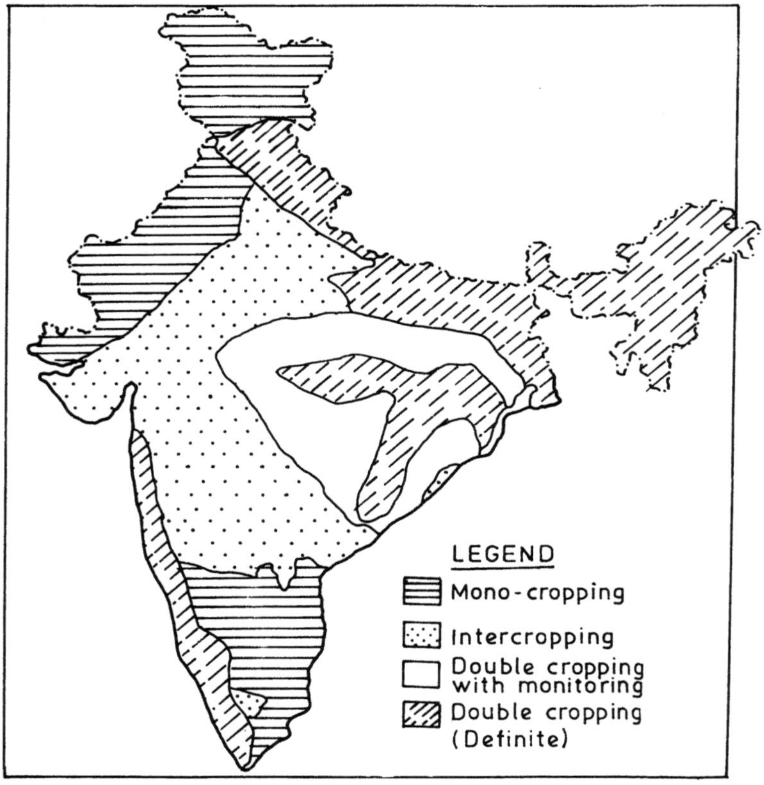

Figure 3. Potential cropping patterns of India. From Venkateswarlu (1985).

type at 75% probability. It would be 20–30 weeks with 25% probability (Table 13).

Taking an overview of all of these situations, India's potential cropping system could broadly be as shown in Figure 3. Simply stated, the guidelines that are listed in Table 14 apply.

Table 14. Guidelines for potential cropping systems in India

Rainfall (mm)	Broad soil group	Growing season (weeks)	Cropping system
350–650	Alfisols and shallow Vertisols; Aridisols and Entisols	<20	Single rainy crop
350–650	Deep Aridisols and Inceptisols	20	Either rainy or postrainy crop
350–650	Deep Vertisols	20	Sole postrainy crop
650–800	Alfisols, Vertisols, Inceptisols, Mollisols	20–30	Intercropping
800 and above	Deep Vertisols, deep Alfisols, Entisols	>30	Double-cropping

Table 15. Crop choice in relation to stored moisture in Vertisols of low-rainfall (508 mm) Bellary in Deccan region

Available moisture (mm)	Choice of crops
>200	Sorghum (125–135 days); safflower (110 days); chickpea (75 days)
150–200	Sorghum (95–105 days); safflower and chickpea
100–150	Safflower, chickpea, sorghum for fodder (75 days)
<100	Dolichos (90 days)

The suitable cropping systems for different agro-climatic regions have been reviewed by several authors (Randhawa and Venkateswarlu, 1980; De and Singh, 1983; Venkateswarlu, 1985). Even in a given region depending on the physiography, the cropping system changes (Randhawa and Venkateswarlu, 1980). For instance, in Bellary region (Vertisols) crops are grown in the postrainy period. The crop choice is to be tailored with the stored water (Table 15). Similarly, in Indore (Vertisols) where the crops can be grown during both the seasons with 990 mm of rainfall, the cropping systems vary depending on the depth of the soil (Table 16). It varies from monocrop in shallow soil to double-cropping in deep soils.

In the Aridisols of northwest India, the rainfall is generally low (380–480 mm). The soils vary from loamy sands to sands. For such soils when depth is not limiting, the texture becomes important. Thus, postrainy crops are suggested based on the rainfall in the monsoon and the soil texture (Table 17). Similarly, for the Inceptisols of Punjab, light soils (loamy sands) can support a rainy crop or postrainy crop. But a sandy loam can support both rainy and postrainy crops.

Table 16. Efficient crops in relation to soil depth in high-rainfall (990 mm) Vertisols of Indore region

Soil depth	Available moisture (mm)	Cropping system
Shallow	100	*Monocropping*: Sorghum, corn
Medium	150	*Monocropping/intercropping*: Soybean, sorghum + pigeonpea
Deep	200	*Double-cropping*: Corn—safflower Soybean—chickpea Sorghum—chickpea

Source: Randhawa and Venkateswarlu (1980).

Table 17. Postrainy crops for Aridisols of northwest India with 450 mm of rainfall

Rainfall in the southwest monsoon (mm)	Appropriate crops for the postrainy period	
	Loamy sands	Sands
<200	*Eruca sativa*	*Eruca sativa*
200–300	Mustard	*Eruca sativa*
300–500	Mustard/chickpea	Mustard
>500	Rapeseed/chickpea	Mustard/chickpea

Source: Venkateswarlu (1981).

V. Tillage

Kanitkar et al. (1968), while reviewing the earlier research efforts on improved crop managements in drylands, described the Madras/Bombay/Deccan Dry Farming methods, which among others included deep plowing (>25 cm) once in 3 years for better intake and storage of precipitation.

Subsequent research suggests that proper tillage is needed for (1) early seedbed preparation for timely sowing, and (2) enabling precipitation to enter the soil. Primary tillage is essentially done by a wooden country plow prior to sowing of crops. It is generally a shallow tillage (15–22.5 cm) through opening of furrows on a criss-cross pattern. This is then followed by blading with a harrow. Besides opening the soil surface, it also helps in minimizing weed problems.

A. Light Cultivation

In light-textured soils, shallow preparatory tillage was found to be beneficial. In the Alfisols of semiarid Telengana (Andhra Pradesh), off-season plowing

Table 18. Effect of off-season tillage on yield of sorghum in semiarid Alfisols (Hyderabad)

Treatment	Grain yield ($Mg\ ha^{-1}$)	
	1978*	1979*
Off-season tillage	2.87	1.24
Direct seeding (control)	2.40	0.46

Source: Venkateswarlu (1980).

* Rainfall (mm)
1978 = 933
1979 = 568

followed by blading was found useful in increasing the yield of crops (Venkateswarlu, 1981). The effect of such tillage is more pronounced in subnormal rainfall years (Table 18). Such a system permits early sowing, which is a very important requirement in drylands, particularly when crops are grown during the rainy season.

Excessive tillage leads to severe soil erosion. As much as 40 t ha^{-1} of surface soil was lost from a field in 6 days during a dust storm (30 km/hr) in 1979 in an Aridisol (loamy sand) at Jodhpur (Gupta and Gupta, 1981).

B. Deep Tillage

Deep tillage is needed at least in three distinct situations. They are:

1. Areas with hard pans (e.g., Inceptisol of Gangetic Plains, which tend to develop hard pans due to consistent cultivation with a country plow)
2. Areas with persistent weeds (e.g., *Cyperus* and *Cynadon*)
3. Areas with textural profiles (Alfisols and related red soils)

Hard pans and soil layers with higher clay due to iluviation have the same problem and that is high bulk density. The red sandy loam soil series, having clayey subsoil (Alfisols) in Andhra Pradesh and Tamil Nadu and black clayey soils (Vertisols) in Andhra Pradesh, and the alluvial sandy loam soil series, belonging to Inceptisols in the north, have such high bulk density layers at shallow depths (Gupta et al., 1984). These layers in the subsoil restrict root growth as was reported in sugarcane (*Saccharum officinarum* L.) (Singh, 1964). Chiseling of such soils to a depth of 30–45 cm at 60–120-cm intervals has been reported to increase the growth and extension of crop roots and grain yield, as well as residual moisture (Reddy and Dakshinamurthy, 1971; Mallick et al., 1978). In black clay soils chiseling up to 20 cm along with mixture of gypsum at 5.0 T ha^{-1} or FYM at 25.0 T ha^{-1} increased sugarcane yield by 16–25% over the control yield.

Deep plowing with moldboard plow up to 25 cm not only improved yields but also had residual effects for at least 2 years at Anantapur in a red sandy

Table 19. Deep tillage (moldboard plow) and its residual effect on yields of various crops in semiarid Alfisol of Anantapur

Crop	Influence in first year (1972–73) ($Mg\ ha^{-1}$)		Residual effects in tillage ($Mg\ ha^{-1}$) 1973–74		1974–75		1975–76	
	S	D	S	D	S	D	S	D
Castor (taproot)	0.64	0.88	1.01	1.30	0.58	0.86	1.26	1.51
Pigeonpea (taproot)	0.90	1.17	0.91	1.18	1.03	1.50	1.52	0.74
Peanut (medium rooted)	0.42	0.54	0.86	1.17	0.57	0.82	1.33	1.42
Pearl millet (shallow rooted)	1.20	0.54	0.77	1.03	0.49	0.74	0.02	0.03
LSD (0.05)	0.24		0.21		0.21		NS	

Source: Anderson (1980).

S = shallow tillage (15–22.5 cm); D = deep tillage (>25 cm); LSD = least significant difference (0.05); NS = not significant.

loam (Table 19). The effect of deep plowing was visible in deep-rooted as well as shallow-rooted systems (Anderson, 1980). Similar results were obtained in Bangalore in a deep red sandy loam. In fact, plowing up to 30 cm increased moisture penetration to as deep as 90 cm, while unplowed land had a wetting front only to 60 cm (Hadimani et al., 1982). On the other hand, deep plowing to 30 cm did not generally lead to increased yields in Vertisols, even though there was improvement in physical parameters like increased infiltration rate and decreased bulk density.

The effect of relatively deep tillage on yield of crops was discussed by Randhawa and Venkateswarlu (1980). They found that the yield increases were more in deep soils growing monsoon crops. Deeper root system crops performed superior over shallow-rooted crops. Vittal et al. (1983) working on an Alfisol of the semiarid tropics region found that deep tillage leads to increased yield of shallow- and deep-rooted crops equally in normal rainfall years. However, the benefit was absent in subnormal rainfall years and was restricted to deep-rooted crops in high rainfall years.

VI. Mulching

Mulching aims at slowing the movement of moisture from soil into the atmosphere. It has been a subject of much interest and has been studied intensively concerning the theory and mechanisms.

Prihar et al. (1979) pointed out that shallow tillage and application of organic mulches at the surface are beneficial to conserve water, particularly in the upper layers of the profile in a loamy sand. It was 0.07 kg per kg of dry soil more with mulching, varying from 0.03 to 0.27 kg kg^{-1} depending on the rainfall. Chaudhary and Sandhu (1982) reviewed the effect of mulches on increasing or decreasing soil temperatures. The mulches advance seedling emergence by 24–34 hours. There would be denser rooting in the top layer due to favorable soil temperature. Furthermore, there would be a moderation in soil temperatures, improving the soil temperatures leading to better growth and improved yields. The mulches are also looked upon as material for reducing soil erosion (Tejwani and Bharadwaj, 1982).

Of the various mulches studied, soil mulch and organic mulch are of importance in the Indian context where crops are grown, both with the monsoon rains as well as during postmonsoon period. And studies on mulching were carried out for both situations.

Randhawa and Venkateswarlu (1980) indicate that mulching has very little place in crop production during the rainy season, but where soil mulch is created as part of weed control practice, it may serve a purpose. However, postrainy season experimental data from many locations revealed that under moisture stress conditions, or when moisture can be carried over for a short time or can be conserved for a subsequent crop, mulching has been beneficial (Table 20).

Moisture conservation in postrainy situations had been favorable (Prihar et al., 1979). On the other hand, there was hardly any difference between no-mulch and soil/straw mulch for pearl millet (*Pennisetum americanum* (L.) Leeke) grown in rainy season (Venkateswarlu et al., 1986). However, straw

Table 20. Effect of mulches on yield of crops

Center	Season	Crop	Yield of grain (Mg ha^{-1}) Control	Mulches
Hoshiraphur (Inceptisol)	Postrainy	Wheat	2.86	3.51
Dehra Dun (Inceptisol)	Postrainy	Wheat	2.33	2.93
Anand (Entisol)	End of rainy	Tobacco	1.33	1.84
Kovilpatti (Vertisol)	Postrainy	Sorghum	0.53	0.94
Varanasi (Entisol)	Postrainy	Barley	1.75	1.91
Solapur (Vertisol)	Postrainy	Sorghum	0.98	1.64

Source: Randhawa and Venkateswarlu (1980).

Table 21. Effect of mulches on moisture depletion by crops in semiarid Alfisols (Hyderabad: 1980–1984)

System	Moisture depletion (mm in 45 cm)	
	Pearl millet	Castor
No mulch	32.6	80.1
Soil mulch	32.8	78.2
Straw mulch	32.8	83.5

Source: Venkateswarlu et al. (1986).

Table 22. Effect of vertical mulching on yield of sorghum (postrainy) in Vertisol of Bellary

Interval	Grain yield ($Mg\ ha^{-1}$)*				
	1973	1974	1975	1976	1977
8 m	0.28	1.61	1.17	1.12	1.92
4 m	0.40	1.69	1.78	1.25	1.54
Control	0.02	1.12	1.10	1.08	1.45

Source: Ramamohan Rao (1981).

* Rainfall in the season (Aug–Dec) in mm for respective years was 207, 554, 474, 570 and 119.

mulch provided more moisture for a long-duration (150 days) castor (*Ricinus communis* L.) crop (Table 21). This was because moisture was carried over for a short period after the cessation of rains for this long-duration cash crop. However, this did not lead to increase in yields. In fact, there was a significant reduction in yield from 1,324 to 1,066 kg ha^{-1} without and with sorghum (*Sorghum bicolor* (L.) Moench) straw mulch. The mulch seems to have caused an allelopathic effect as earlier reported by Guenzi et al. (1967).

A. Vertical Mulching

Keeping sorghum stubbles as vertical mulch in trenches (40 cm deep, 15 cm wide, and protruding 10 cm above the ground level) enhanced the available soil moisture by 4–5 cm and improved grain yields by 40–50% in normal years and by as much as 400–500% over unmulched plots in below-normal rainfall years in a Vertisol at Bellary (Table 22).

Thus the effect of mulching, by and large, is beneficial to postrainy crops in India. This also enables the farmers to use most of the mulch back as a feed as it does not tend to decompose faster in the postrainy period. And this is so important since most of the organic residues (straw, haulms) are also used as feed in India where there is a great need because of bovines.

Table 23. Effect of good farming practices on the yield of sorghum

System	Hyderabad (Alfisol)—Poor in fertility (Mg ha^{-1})	Bellary (Vertisol)—Moderate in fertility (Mg ha^{-1})
Farmers traditional practices	0.47	0.74
Improved farming practices	0.77	1.39
Traditional practices + fertilizer (40–13.1–0)	1.11	1.65
Improved practices + fertilizer (40–13.1–0)	1.98	2.75

Source: Venkateswarlu (1985).

VII. Fertilizer Use

Drylands are sometimes more hungry than thirsty. The soils in India are universally deficient in nitrogen. Phosphorus is low in 43.9% of soils (Venkateswarlu, 1986). It is found to be limiting at high production levels (Finck and Venkateswarlu, 1982). Potash is adequate in many areas except in light-textured soils or at continued high levels of crop production.

Inter alia yield is dependent on soil, inputs, crop (variety), climate, and management. In other words, proper soil and crop husbandry are prerequisites to using external inputs as critical supplements. The data in Table 23 illustrate this point. In the Alfisols of Hyderabd, the inherent fertility is poor as is evident from the control yield. With the use of 53 kg of nutrients per ha over the existing farmers' management systems, the response was 12.0 kg of grain per kg of nutrient. On the other hand, when the improved practices (discussed elsewhere) were superimposed, the response increased to 28.4 kg of grain per kg of nutrient. In Bellary with a Vertisol, the native fertility being superior, the response with farmers and improved management systems worked out to 17.1 kg and 37.8 kg per kg of nutrients, respectively.

The message is clear. If fertilizer use is to be effective, improved management systems must be introduced.

The cardinal principles in fertilizer use are:

1. Law of Diminishing Returns
2. Law of Minimum

The response of sorghum CSH-6 in a Vertisol and in an Alfisol of ICRISAT are given in Figures 4 and 5, respectively.

A perusal brings out:

1. The response to the first increment is always higher irrespective of the rainfall received during the growing season.
2. As the Vertisols have more carry-over moisture, there is stability in production. Yields up to 5.0 Mg ha^{-1} were achieved even in low moisture

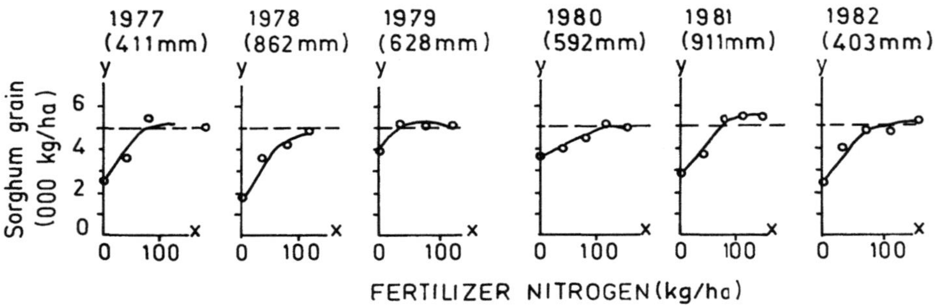

Figure 4. Yield of sorghum in Vertisols.

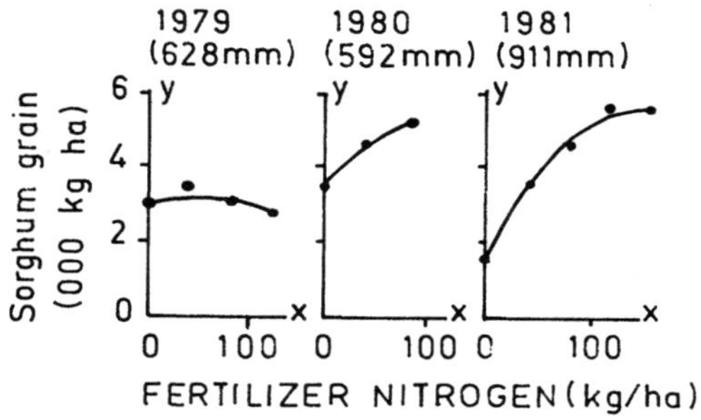

Figure 5. Yield of sorghum in Alfisols. From ICRISAT (1984).

Table 24. Relative response of crops to moderate levels of nitrogen

Crop	Fertilizer level (kg N ha^{-1})	Average response (kg grain kg^{-1} N)
Rainy season crops		
Pearl millet	40	19.2
Sorghum	40	26.7
Corn	40	15.5
Rice	40	20.3
Postrainy crops		
Sorghum	30	14.4
Barley	20	22.0
Wheat	25	31.0
Safflower	25	15.8
Mustard	20	13.6

Source: Venkateswarlu (1980).

Table 25. Relative response of crops to phosphorus at 13.1–21.8 kg ha^{-1}

Crop	Average response to P (kg grain kg^{-1} P)
Sorghum	20.6
Pearl millet	16.5
Corn	27.9
Safflower	7.3
Mustard	13.0
Chickpea	25.0

Source: Venkateswarlu (1980).

Table 26. Effect of nitrogen and phosphorus on yield of crops

			Average yield (Mg ha^{-1})			
Center	Soil type	Crop	Control	N alone	P alone	N + P
Hyderabad	Alfisol	Sorghum	0.67	0.78	1.16	2.24
Bangalore	Alfisol	Finger millet	1.76	2.15	1.93	3.06
Rajkot	Vertisol	Sorghum	0.39	0.40	0.22	0.92

Source: Venkateswarlu (1980).

retentive Alfisols where the clay was varying from 15% to 30% depth being moderate.

An examination of the response data to nitrogen and phosphorus for various crops in the AICRPDA show that they are comparable to, if not better than, the irrigated crops (Tables 24 and 25). The response to phosphorus is less conspicuous. The fertilizer recommendations for drylands, therefore, need to be at moderate levels.

Response to highly deficient nitrogen can be low if other nutrients are limiting. More frequently, phosphorus is limiting the response to applied nitrogen in the semiarid Alfisols as well as Vertisols (Table 26).

However, Venkateswarlu (1986) pointed out that the need for phosphorus occurred only at increased productivity levels. The more conspicuous results are observed with shallow-rooted (cereal) crops grown on Alfisols. The response was relatively less on Vertisols, deep submontane soils (Mollisols, Inceptisols, Entisols), and alluviums (Inceptisols, Alfisols, and Entisols). Furthermore, deep-rooted crops and crops with higher exchange capacity of roots are found to feed more efficiently on native phosphorus, including the difficulty of soluble phosphorus Mycorrhizal associations are found helpful for increasing nutrient availability, including the less available forms of phosphorus.

With potassium, the responses are less frequent. The native available and fixed-potassium (molar 1.0 HNO_3 soluble) are fairly adequate in many soils. There have been long-term potassium experiments in the country. Venka-

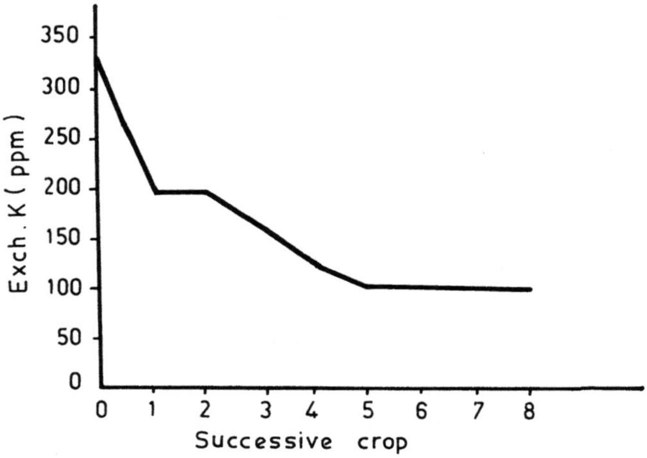

Figure 6. Depletion patterns for exchange potassium in heavy black soils. From Venkatasubbaiah et al. (1976).

Table 27. Inputs and removal of potassium from an Alfisol in the long-term potassium experiment, ICRISAT (1979–1983)

Inputs of K to soil ($kg\ ha^{-1}$) 1979–83	Treatments				
	Control	N_{120}	K_{30}	K_{60}	K_{120}
Removal in herbage	221	252	302	327	355
Added	—	—	170	240	480
Net input	−221	−252	−182	−87	+125
Exchangeable K in soils (ppm)					
1979	65	54	60	62	61
1983	49	48	57	72	84
Change in 4 years	−17	−4	−6	−3	+10
Source of K removal					
Exchangeable K	−71	−25	−13	+42	+101
Nonexchangeable K	−150	−227	−169	129	+24

Source: International Crops Research Institute for Semi-Arid Tropics (ICRISAT) (1984).
Note: Basal application of 120 kg N ha^{-1} over K treatments and 15 kg P ha^{-1} over all treatments; Calculated from exchangeable K concentration by assuming soil apparent density = 1.4 g/cc.

tasubbaiah et al. (1976) in a potted plant study found that the depletion of potassium in Vertisols was as shown in Figure 6. The stabilization was at 100 ppm, which is fairly adequate for years to come. At ICRISAT in a long-term trial, response to potassium was more from the 3rd year onward in an Alfisol with sorghum. The inputs and removal of potassium from the Alfisol in the long-term study are given in Table 27. In another potted plant study

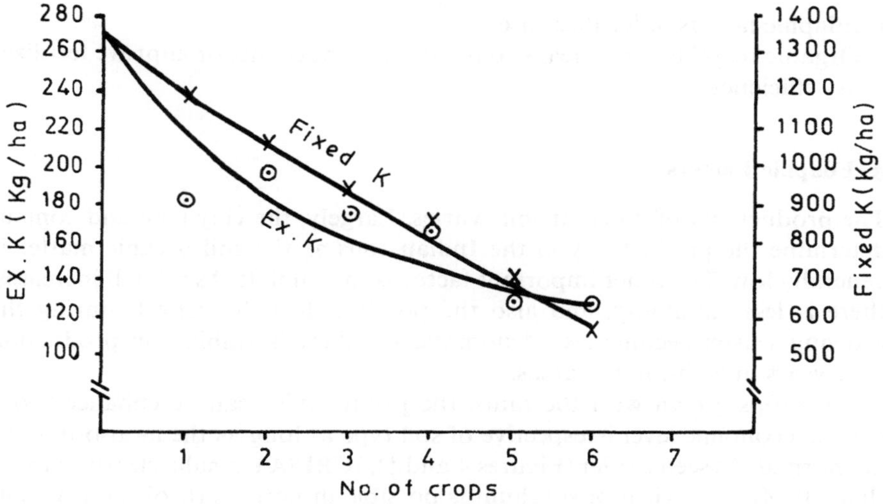

Figure 7. Potassium-depletion pattern with pearl millet (pot study). From Venkateswarlu (1984b).

(Venkateswarlu, 1984b), the potassium-depletion pattern with pearl millet was as given in Figure 7. These studies suggest that under high production systems and in light-textured soils, the need for potassium may be significant.

Responses to calcium, sulfur, and micronutrients have been inconsistent. In the intensive peanut-cultivated areas of Gujarat, phosphate-induced sulfur deficiency has been reported in peanuts (D.P. Motiramani, personal communication, 1984). That increased phosphorus availability reduced the availability of sulfur was earlier reported in a pot study (Venkateswarlu and Subbiah, 1964). Zinc deficiency was more frequent at high production levels as well as in light-textured soils (e.g., Alfisols) with higher rainfall. Calcium-deficient areas were found most often in peanut-growing areas, particularly in light-textured soils (Singh and Venkateswarlu, 1985). Thus the limiting nutrients, other than nitrogen and in many instances phosphorus, are site specific. But oftentimes, blanket recommendations are made, overemphasising the need for balanced fertilization. This approach is *not* advisable for drylands where the farmers are generally poor in resources. External inputs must be added only when the response to the same is assured, since the risk in crop production is more in drylands.

The fertilizer use efficiency can be increased by placement of a basal dose of nutrients and by split application of nitrogen to monsoon crops. The fertilizer for postrainy crops must be deep placed in the moist zone for increasing its use efficiency.

In the recent past, two areas of fertilizer use research have received attention. They were:

1. Edaphic factors in fertilizer use
2. Organic recycling as a means to incorporate nutrients or improve fertilizer use efficiency

A. Edaphic Factors

The productivity of different soils varies. Largely, the clay type and content determine the productivity in the Indian soils as the soil organic matter is generally low. The other important factor is the rainfall. As rainfall increases, there is less variability, and also the possible drought periods during the growing season become less. Axiomatically, there is stability in production over years in high-rainfall areas.

For crops grown with the rains, the productivity can be enhanced to a desired economic level irrespective of soil type as long as the monsoon rains are normal. As seen earlier (Figures 4 and 5), ICRISAT results clearly showed that 5.0 Mg ha^{-1} yield of sorghum is possible in both Vertisols and Alfisols unless the rainfall is too inadequate, as in the year 1979.

However, the productivity of the postrainy crops varies significantly with the soil depth and/or texture. In Vertisols, it is the depth that determines the plant-available water, while in Entisols and Inceptisols, it is the texture that is most important.

Umrani (1980) has shown that in Vertisols as the depth increases the productivity increases (Table 28) for a given fertilizer schedule. On the other hand, Singh et al. (1975) showed for Inceptisols (submontane region) that productivity increased as the soil texture became finer (Table 29).

B. Organic Recycling

The use of feedyard manure (FYM) to dryland crops is a common practice in India. However, it is largely practiced only near villages. This is primarily because of two factors. One is that most of the farmers have some parcels of land near villages that can be guarded and cared for. The other is the logistics.

Table 28. Effect of soil depth on yield of crops at a given fertilizer level in Vertisols of Solapur

		Yield of crops (Mg ha^{-1})		
Soil depth (cm)	Stored water (mm)	Sorghum (50–0–0)	Safflower (50–10.9–0)	Chickpea (50–10.9–0)
30	9	1.29	0.35	0.62
60	18	2.13	0.49	0.71
90	27	—	1.05	1.19

Source: Venkateswarlu and Singh (1982).

Table 29. Effect of soil texture on yield of wheat in Inceptisol soils of Hoshiarpur

Texture	Stored water (m^3 water m^{-3} dry soil)	Yield (Mg ha^{-1})
Loamy sand	0.134	2.41
Sandy loam	0.150	3.55
Clay loam	0.246	4.16

Source: Singh et al. (1975).

It is difficult to carry FYM long distances. Generally speaking, the better lands are nearer to the villages and the richer communities also possess most of these lands. As a corollary then, the poorer communities use less of the available FYM and their lands are also of poorer endowments. Hence, there has been a conscientious effort to identify measures of on-farm residue management.

Green manuring of dryland crops is rare, but practiced in small pockets in India. For instance, in the Khammam district of Andhra Pradesh, mungbean (*Vigna radiata* (L.) Wilczek) is taken as an early crop with June rains and by late August to early September it is harvested and the residue is incorporated. Within a fortnight, sorghum is seeded. These soils are generally Vertisols and receive higher rainfall (>1,000 mm). Sorghum is their important crop. So, if sowing of mungbean is delayed, then they would harvest the mungbean at physiological maturity by pulling out the whole plants, prepare the land, and sow the next crop, (i.e., sorghum).

The effect of green manuring and FYM on yield of crops is primarily due to the supply of nutrients, but also to an improved physical environment. The effect on yield is given in Table 30, along with fertilizer equivalence.

1. Residue Incorporation

In the Alfisols receiving 924 mm (Bangalore) and 760 mm of precipitation (Hyderabad), on-farm residue incorporation at 4.0 Mg per ton of maize (Zea mays) stalks and 2.0 Mg per ha of cowpea (*Vigna unquicultа* (L.) Walp.) residue was tested over 5–6 years on maize and pearl millet (Tables 31 and 32). The results duly show the advantage of on-farm residue management. As reported elsewhere, the physical environment also improved with such an incorporation. Such incorporation not only improved the yield, but also improved the physical environment, as well as the soil fertility status (Table 33). Similar results were reported by Lockeretz et al. (1981) while comparing organic farming with conventional farming in the corn belt of the United States.

Organic matter with a low C:N ratio (<75) decomposes very fast, and has a rapid and positive effect on aggregation. On the other hand, organic matter with a high C:N ratio (>150) decomposes too slowly to be effective. The most suitable C:N ratio has been found to be 75–150. Most of the crop residues

Table 30. Effect of Feedyard manure and green manure on yield of crops

Center	Soil type	Crop	Rainfall (mm)	Average yield (mg ha^{-1})			
				Control	GM + fertilizer	FYM + fertilizer	Recommended fertilizer
Bangalore	Alfisol	Finger millet	924	1.23	2.15	2.32	2.18
Hyderabad	Alfisol	Castor	760	0.42	0.72	—	0.78
Jodhpur	Aridisol	Pearl millet	380	0.88	—	1.54	1.38
			Fertilizer		*Fertilizer*		*GM or FYM (Mg ha^{-1})*
Bangalore			25–10.9–20.8		50–21.8–41.7		10
Hyderabad			10–13.1–0		40–13.1–0		4
Jodhpur			10–0–0		40–0–0–		10
Fertilizer equivalence per Mg of GM/FYM as kg N ha^{-1}							
Dryland				*FYM*		*GM*	
Bangalore	Alfisol	Finger millet		7.0		5.8	
Hyderabad	Alfisol	Castor		—		4.9	
Jodhpur	Aridisol	Pearl millet		4.0		—	
Irrigated							
Hyderabad	Vertisol	Rice		5.0			
Ludhiana	Inceptisol	Corn		6.8			
Bangalore	Alfisol	Finger millet		7.1			

Source: Annual Report of the Long Term Fertilizer Project for 1979–1980 (ICAR); Annual Reports of AICRPDA 1973–1978 of Bangalore, Hyderabad, and Jodhpur.

FYM = feedyard manure; GM = green manure.

Table 31. Effect of continuous application of corn residue at 4 Mg ha^{-1} on the yield of corn grown on Alfisol near Bangalore

	Yield of corn (Mg ha^{-1})[a]	
Year	With residue	Without residue
1975	3.57	2.95
1976	1.15	0.75
1977	3.56	2.80
1978	2.52	2.23
1979	0.99	0.69
Average	2.35	1.88

Source: Hadimani et al. (1982).

[a] Basal dose of 50–21.8–41.7 was applied.

Table 32. Effect of continued cowpea residue incorporation on the yield of pearl millet in Alfisol of Hyderabad

		Yield (Mg ha^{-1})	
Year	Rainfall (mm)	With residue	Without residue
1974	507	1.58	1.08
1975	724	Failed[a]	Failed
1976	366	1.48	1.40
1977	348	2.05	2.01
1978	578	2.04	2.01
1979	227	Failed	Failed
1980	332	2.43	2.13
1981	501	2.08	1.88
Average		1.93	1.75

Note: Crop failed in 1975 due to downy mildew.

Source: Venkateswarlu (1984b).

[a] Residue added at 20 kg N ha^{-1} equivalence.

Table 33. Effect of organic recycling[a] on the chemical properties in Alfisol of Hyderabad

	Loamy sand		Sandy loam	
Soil constituent	Initial	After 5 years	Initial	After 5 years
Organic carbon (%)	0.34	0.39	0.65	0.62
Available N (kg ha^{-1})	235	286	284	316
Available P (kg ha^{-1})	2.9	5.6	2.4	6.4
Available K (kg ha^{-1})	224	214	213	215

Source: Venkateswarlu (1984b).

[a] Residue of cowpea and pearl millet in rotation incorporated equivalent to 20 kg N ha^{-1}.

Table 34. Potential areas for adoption of green manuring in drylands

Region	Soil type	Rainfall (mm)	Cropping system with possibility of organic recycling
Bangalore	Alfisol	924	Cowpea-finger millet
Bhubaneswar	Alfisol	1,463	Mungbean-finger millet
Rewa	Vertisol	1,168	Blackgram-wheat
Agra	Inceptisol	765	Mungbean-mustard
Akola	Vertisol	877	Mungbean-safflower
Varanasi	Entisol	1,113	Rice-chickpea
Hoshiarpur	Inceptisol	1,000	Maize-chickpea
Indore	Vertisol	1,053	Maize-chickpea
Ranchi	Alfisol	1,462	Ragi-chickpea
Dehra Dun	Inceptisol	2,313	Maize-chickpea

Note: In the first 5 years, the residue of rainy crop is to be incorporated, and in the rest, the residue of postrainy crop is to be incorporated.

fall in this range. If not, inorganic nitrogen can be added to hasten the process of decomposition. Such potentialities exist in several parts of India (Table 34).

2. Improving Soil Productivity Through Cropping Systems

India is a vast country with a large variety of crops and agro-climatic situations. Intercropping and double-cropping is possible. Food legumes and fodder legumes can find a place in these systems. The nitrogen economy can be improved with such an approach. Even when a legume is rotated with a cereal, there is a definite improvement in the yield of cereal. In fact, Smith (1942) suggested that there is increased fertilizer nitrogen use efficiency with a legume being a part of the crop rotation. Some results citing Indian experience are discussed in the following paragraphs.

a. Legume in Crop Rotation

In low-rainfall areas (<750 mm), only sole cropping is generally suggested. At the most, there could be intercropping. In such systems, a legume either as a food or fodder can be introduced. For instance, on an Alfisol near Hyderabad (760 mm) and on an Inceptisol near New Delhi (714 mm), such a rotation benefited the cereal crop (Tables 35 and 36).

In Hyderabad, a perceptible increase in yield of sorghum, with peanut, mungbean, and cowpea as the preceding crop, was obtained. Similarly, in Delhi the effects were observed with pearl millet (Giri and De, 1979). The nitrogen uptake increased from 29.5 kg ha^{-1} to 39.9 kg ha^{-1}, on an average, with the legumes.

A fodder legume (*Stylosanthes hamata*) can be grown in rotation with cereals. The legume effect of the fodder legume is quite visible (Table 37). These

Table 35. Effect of previous crop in rotation on sorghum in Alfisol of Hyderabad

Previous crop	Yield of sorghum (Mg ha^{-1}) at different levels of N (kg ha^{-1}) 0	20	40	60	Average
Peanut	1.70	2.87	2.44	3.21	2.55
Mungbean	2.26	2.86	2.63	2.74	2.62
Cowpea	2.52	2.48	2.90	3.03	2.73
Pigeonpea	1.84	1.92	2.13	2.22	2.03
Sorghum + pigeonpea	1.73	1.93	2.44	2.22	2.08
Castor	2.03	2.09	2.44	2.13	2.10

Source: Singh and Das (1984).

Table 36. Effect of previous crop in rotation on the pearl millet in Inceptisol of Delhi

Previous crop	Rate of N (kg ha^{-1}) applied to pearl millet (Mg ha^{-1}) 0	30	60	Average
Mungbean	3.08	3.61	3.58	3.42
Peanut	3.03	3.72	3.72	3.62
Cowpea	3.75	3.97	3.94	3.89
Pigeonpea	3.00	3.68	3.65	3.44
Pearl millet	2.55	3.11	3.60	3.09

Source: Giri and De (1979).

Table 37. Yield of sorghum and pearl millet after *Stylosanthes* in rotation (Alfisol: Hyderabad, 1981)

	Grain yield (Mg ha^{-1})			
	Sorghum		Pearl millet	
N level (kg ha^{-1})	After castor	After *Stylosanthes*	After castor	After *Stylosanthes*
0	0.21	1.08	0.63	2.42
25	0.85	1.18	1.17	2.37
50	1.16	1.20	1.70	2.54
100	1.65	1.74	2.35	2.74
Mean	0.95	1.20	1.46	2.51
LSD	0.26		0.32	

Source: Venkat Reddy and Hampaiah (1982).

LSD = Least Significant Difference

results have relevance in the Indian context, since even the farm holdings of small and marginal farmers would be scattered in several places, as in the case of big farmers. And more frequently, one or more parcels of land would be left fallow due to lack of time or resources. In such situations, the fallow lands can be put under *Stylosanthes hamata* for improving the soil productivity and also for producing nutritious legume fodder.

b. Legume in Intercropping

The legume in intercropping could be short (*Vigna* group) or long (Pigeonpea—*Cajanus cajan* (L.) Millsp.) in duration. Both have advantages for the associate nonlegume crop. The *Vigna* group is harvested as the earlier crop and provides additional income to the farmer besides leaving nitrogen for use by the longer duration companion crop. On the other hand, pigeonpea is least competitive during the growth and development and after the harvest of the companion crop; it starts using the space by putting out additional growth. By the time development starts, leaf shedding occurs, leading to an addition of organic nitrogen up to 50 kg ha^{-1} (Pratap et al., 1980) to the crops of the next season grown in a Vertisol. Similar results were reported by Kumara Rao et al. (1983) on maize after pigeonpea in a Vertisol.

Some of the results are given in Tables 38 and 39. It is obvious that in all instances the associate nonlegume is benefited. With cowpea, sorghum had more benefit when the crop was harvested as a fodder, both in terms of yields as well as nitrogen uptake. It could be argued that this might be due to early clearance of the land. But earlier work of Venkateswarlu (1985) in a potted study proved that with the soil after cowpea as fodder the uptake of nitrogen was 40.3 mg per pot by foxtail millet (*Setaria italica* Beauv.), as compared with the soils after cowpea as grain when the uptake was only 35.0 mg per pot. The mungbean and cowpea gave an estimated advantage of 30 kg N ha^{-1} (Singh and Das, 1984) for sunflower (*Helianthus annuus* L.) grown in an Aridisol at Jodhpur.

Table 38. Yield of sorghum under different intercropping systems of sorghum:cowpea (1:1) in Alfisol of Hyderabad

System	Fertilizer used (kg ha^{-1})	Yield (Mg ha^{-1})	Uptake of N (kg ha^{-1})
Cowpea (grain)	0–13.1	1.93	53.7
Cowpea (grain)	30–13.1	2.35	57.8
Cowpea (fodder)	0–13.1	2.74	65.5
Cowpea (fodder)	30–13.1	3.10	77.4
Without cowpea	30–13.1	3.30	69.8
LSD (0.05)		0.66	12.90

Source: Venkateswarlu (1984b).

LSD = Least Significant Difference

Table 39. Yield of sunflower under different intercropping systems of crops in Aridisol of Jodhpur

System	Yield (Mg ha^{-1}) 0	30 (N Kg ha^{-1})	60
Pearl millet	0.31	0.52	0.80
Mungbean	0.13	—	—
Cowpea	0.13	—	—
Sunflower + mungbean	0.16 + 0.10 (1.28)	0.48 + 0.11 (1.75)	0.66 + 0.11 (1.66)
Sunflower + cowpea	0.14 + 0.93 (1.18)	0.51 + 0.10 (1.80)	0.66 + 0.11 (1.73)

Note: Figures in parentheses give Land Equivalent Ratio values.
Source: Singh and Das (1984).

Table 40. Effect of different management systems of alley cropping with *Leucaena leucocephala* on yield of postrainy sorghum grown in Vertisol of Solapur

System	Grain yield (Mg ha^{-1})	Total N uptake (kg ha^{-1})[a]
No *Leucaena*; no loppings added	0.62	16.64
No *Leucaena*; loppings added	1.07	26.27
Leucaena; no loppings added	0.48	13.47
Leucaena; loppings added	0.93	24.34

Source: Singh and Das (1984).
[a] Composite samples.

Where alley cropping is practiced with *Leucaena leucocephala*, there is yield advantage for postrainy sorghum in Vertisols of Solapur (Table 40). Also, the uptake of nitrogen considerably improved. The tree component affects the yield of the associate sorghum crop. It could, however, be compensated with addition of the loppings of the tree component. Similar results were obtained at the International Institute of Tropical Agriculture (IITA) in Nigeria (Kang et al., 1981).

c. Legume in Double-Cropping

In high-rainfall situations (800 mm and above), double-cropping is possible and one of the crops can be a legume. Some of the results in the Entisols of Varanasi and Inceptisols of Delhi are presented in Tables 41 and 42. In Delhi the effects of legumes are more spectacular than at Varanasi.

To sum up, then, organic recycling in any form should be encourage for it

Table 41. Effect of preceding crops and nitrogen does rate on the grain yield of crops (Mg ha^{-1}) in Entisol of Varanasi

Nitrogen rates[a]	Grain yield (Mg ha^{-1}): Corn as preceding crop (80 kg N + 17.5 kg P ha^{-1})			Blackgram as preceding crop (17.5 kg P ha^{-1})		
	Mustard	Wheat	Barley	Mustard	Wheat	Barley
1	0.54	1.40	0.86	0.62	1.52	0.5
2	0.67	1.83	1.24	0.6	1.88	1.23
3	0.7	1.83	1.44	0.7	2.14	1.67
4	0.9	1.69	1.76	0.8	1.78	2.49
LSD (0.05)	0.20	0.25	0.31	NS	NS	0.55

Source: Singh and Das (1984).

LSD = Least Significant Difference; NS = not significant.

[a] *N rates*	*Mustard*	*Wheat*	*Barley*
1	0	0	0
2	20	30	20
3	40	60	40
4	60	90	60

Table 42. Legume-mustard double-cropping on yield of mustard in Inceptisol of Delhi

Rainy season crop	Yield of rainy season crop	Rate of N (kg ha^{-1}) applied to mustard (Mg ha^{-1})			
		0	20	40	Mean
Blackgram	1.36	2.15	2.47	2.84	2.49
Mungbean	0.65	1.60	1.92	1.95	1.82
Fallow	—	1.21	1.41	1.79	1.47

Source: Giri and De (1979).

improves the soil's physical and chemical environment. It avoids the problem of transport that is necessary with FYM. On-farm management is easy. A legume would add more nitrogen to the system either when incorporated or when grown along with or before any nonlegume crop. The benefits could vary from 20 to 50 kg N ha^{-1}.

Subsequently, efforts have been made to assess the significance of the nonsymbiotic nitrogen fixers like Azospirillum and Azotobacter. Significant increases in grain yield of pearl millet were reported by Subba Rao et al. (1982). Wani et al. (1985) reported field studies with pearl millet cultivars inoculated with different nitrogen-fixing bacteria. Significant interactions were observed between host cultivars and bacterial strains, but some cultivars showed consistently increased grain and dry matter yields, suggesting the possibility of exploiting suitable plant and nitrogen-fixing bacterial associations for increasing grain yield. However, these results are not consistent and large-scale exploitation becomes possible only when there is assured increase in grain yield of crops. Perhaps what is now needed is to identify the environmental requirements for obtaining increased yields with such nitrogen-fixing bacteria.

VIII. On-Farm Rainwater Management

Rainfall is the only source of water to crop plants grown in dryland areas. The precipitation enters the soil through the surface of the land. If it is not receptive, more of the rainwater moves off the field as runoff. For instance, even though the light soils (loamy sands and sandy loams) have high infiltration, if they form a crust on the surface, more of the rainwater moves off the fields as runoff. Similarly, the deep black soils are characterized by low infiltration. On these lands, the soil surface has to be kept more receptive. There are several measures. Among others, they include:

1. Terracing
2. Interterrace treatment
3. Chemical amendments
4. Crop residue management

A. Terracing

Essentially, the technology includes mechanized structures as shown in Table 43. These structures are meant not only for moisture conservation, but also to minimize loss of fertile surface soil by erosion. The net benefit accruing out of this programme is not beyond 15% additional yield over a base yield of 200–400 kg ha^{-1}, which is not significant enough to warrant adoption by farmers. Dhruvanarayana and Ram Babu (1983) estimated the annual soil loss from water erosion in India. Updating the data on soil loss from black soils, Venkateswarlu (1984a) estimated that the total soil loss is 2,312.1 Mt. Assuming 5.0 Mg ha^{-1} soil loss is tolerable, at least 58.08 Mha of arable land and 6.31 Mha of forest area are subject to greater soil loss and need immediate attention.

Table 43. Mechanical structures of terracing

Mechanical structures	Soil type	Rainfall(mm)	Slope(%)
Contour bunds	Light soils	<600	>1.5
Graded bunds	All soils	>600	1.5–6.0
Bench terraces	Deep soils	>1,000	6–30
Graded border strip	Deep Alfisols and related red soils	>800	>1.5

Thus, these mechanical structures are important in preventing soil erosion. In the recent past, there has been great attention to viewing this problem in an integrated way. One way to save the surface from the beating action of the rain is by providing a vegetative cover (Hudson, 1971). Another approach is to provide vegetative barriers at set intervals instead of terraces. Alley cropping corn (*Zea mays*) and *Leucaena leucocephala* is one such system where the soil erosion can be reduced (Oakes, 1968; Brewbaker and Hilton, 1979). In fragile low-rainfall Sierozems of western Rajasthan, wind breaks and shelterbelts, permanent grasslands in marginal lands, and stubble mulching have been suggested for minimizing wind erosion (Singh, 1982).

Table 44. Useful land configurations for improving yield of crops

Area	Rainfall	System of land configuration	Reference
Alfisols (semiarid)	Medium	Showing across the slope and ridging the land between the rows with later interculturing operations	Shriniwas et al. (1981)
Alfisols (semiarid)	Medium	Graded border strips of 130 × 11 m with 0.1% grade	Dixit et al. (1981)
Alfisols (semiarid) (shallow)	Low	Dead furrows at 2.4–3.6-m intervals	Venkateswarlu (1980)
Vertisols (semiarid or subhumid)	High	Broad bed and furrow with a bed of 100 cm and 50 cm of shallow furrow with a depth of 15 cm	Kanwar et al. (1982)
Vertisols (subhumid)	High	Graded furrows at a grade of 0.2–0.3% and with a cross-section of 0.1–0.15 m^2 and spacing of 8–10-m intervals	Verma (1982)
Vertisols (semiarid)	Low	Contour farming	Ramamohan Rao (1981)
Inceptisols (subhumid)	High	Interplot water harvesting of 1:1, 2:1, or 3:1 of corn:rice areas	Vijayalakshmi (1982)
Aridisols (semiarid)	Low	Interplot water harvesting of 1:1 of cropped to uncropped land, the slope of uncropped slope being on both sides	Singh et al. (1981)
Aridisols (semiarid)	Low	Interrow water harvesting with a treatment of ridges with runoff-inducing substances like tank silt or polythene	Singh et al. (1981)

B. Interterrace Treatment

Even though bund terraces are formed, oftentimes the moisture retention in the interterrace area is not uniform. And it is fairly understood that differential water availability leads to differential responses of the crop to various inputs. In order to even out these differences, certain interterrace treatments are suggested. The aims of these treatments are:

1. To have uniform and on-farm rainwater recharge in the profile
2. To minimize movement of soil within the terraces
3. To increase moisture retention for extended times to either delay drought affects due to short dry spells or to help in establishing a second crop in high-rainfall situations.

The various on-farm land configurations found useful are listed in Table 44. But all of these land configurations are site specific and gave a yield advantage of about 20 ± 5% over control at any given level of productivity. Some of the data given in Table 45 illustrate the point.

C. Chemical Amendments

The ill effects of sodium on soil productivity are well established. On wetting, the sodic soils disperse, and they shrink heavily on drying. The hydraulic

Table 45. Effect of land configuration on yield of crops

Crop	System	Average yield (Mg ha^{-1})	
		Control	With land configuration
Sorghum + pigeonpea (Hyderabad-Alfisols)	Flat on grade and ridging later	2.45 + 0.58	2.91 + 0.72
Finger millet (Bangalore-Alfisols)	Graded border strip with 0.1% grade	2.28	3.35
Peanut (Anantapur-Alfisols)	Dead furrows at 3.0 m	0.68	0.75
Sorghum (Akola-Vertisols)	Broad bed and furrow	1.69	2.13
Sorghum (Bellary-Vertisols)	Graded farming	0.21	0.28
Corn + rice (2:1) (Dehra Dun-Entisols)	Interplot water harvesting	0.97 + 1.0	1.14 + 1.21
Pearl millet (Jodhpur-Aridisols)	Interplot water harvesting (1:1) slope on both sides	1.85	2.80
Pearl millet (Jodhpur-Aridisols)	Interrow water harvesting (polythene)	0.56	0.84

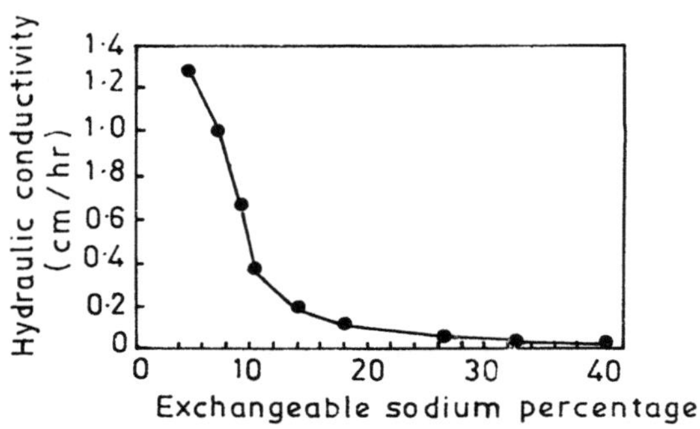

Figure 8. Hydraulic conductivity as influenced by exchangeable sodium percentage. From Ramamohan Rao and Seshachalam (1976).

Table 46. Effect of gypsum application on infiltration rate

	Infiltration rate (mm/hr)				
Treatment	1974	1975	1976	1978	1979
Gypsum	4.1	4.6	4.5	5.5	7.2
No gypsum	0.75	0.75	0.75	1.0	1.0

Source: Ramamohan Rao and Seshachalam (1976).

conductivity is adversely affected with increase in exchangeable sodium percentage (ESP) (Figure 8). Even an ESP of 7.0 becomes critical for Vertisols (Ramamohan Rao and Seshachalam, 1976). Application of gypsum to reduce the ESP to less than 7.0 improved the infiltration rate from 9.9 to 16.5 mm hr^{-1}. For economic consideration it is adequate to supply gypsum to reduce the ESP to about 5.0 and only in the first 10 cm depth (Table 46). There is seldom any indication of the possible influence of exchangeable magnesium on the structure of Vertisols. For example, the plasticity indices of calcium-montmorillonites and magnesium-montmorillonites are very similar (Baver et al., 1977). However, N.S. Jayaraman (personal communication 1983) found synergistic effects of magnesium and sodium in determining the extent of dispersion in Vertisols. This, then, suggests that even at as low an ESP as 3.0, the Vertisols may disperse if the exchangeable magnesium is above 20%. Such situations occur in Gatta Prabha and Mala Prabha areas of the Karnataka north Vertisol region.

D. Residue Management

In India, the stalks or haulms are oftentimes used as fodder or fuel. The estimated yield of residue is 185.3 Mt per annum from the important arable crops (Bharadwaj, 1981). They contribute about 1.24 Mt of nitrogen, 0.16 Mt of phosphorus, and 2.0 Mt of potassium (3.4 Mt of total nutrients). Of this, he estimated that one third would be available for incorporation as residues in arable lands. With the increased scope of having biomass through better management practices, it should not be possible to utilize on the land a part of the residues from the crops or through the trees/shrubs grown in alley cropping systems. On-farm residue management is feasible as it does not warrant transport. Such a system permits leaving the residues as mulch-cum-manure or as incorporated into the soil.

Incorporation of organic matter in any form not only adds nutrients but also improves soil physical conditions. But certain basic principles are important. In the tropics, the rate of decomposition of added organics is so rapid that nearly 90% of what is added is decomposed within 1 year (Joffe, 1955). Then the approach needs to be one of the turnover of organic matter in small quantities, rather than application of large quantities at one time aiming at a buildup at one time. Of all the possibilities, the most attractive one is raising good crops with proper fertilizer use where there would be adequate residues partly for fodder/fuel and also incorporation. On-farm management of crop residues help in avoiding transport problems, more so in hinter lands. If the residues have a wider C:N ratio (>150), a small addition of chemical nitrogen (say 10 kg ha^{-1}) would hasten the process of decomposition and lead to the attendant advantage of improving the soil physical conditions (Krishnamoorthy and Venkateswarlu, 1976).

When 20 kg of nitrogen-equivalent pearl millet or cowpea residue were incorporated in a loamy sand or sandy loam belonging to Alfisols in Hyderabad, the physical environment improved (Table 47). Similar results were obtained by Hadimani et al. (1982) for the Alfisols of Bangalore region where 4 t per ha of corn residue were incorporated annually. The yield increased from 1.88 Mg ha^{-1} to 2.35 Mg ha^{-1} over a 5-year period and the organic carbon increased from 0.55% to 0.90%.

Table 47. Soil characteristics prior to and after crop residue management for 5 years (1974–1978) in an Alfisol of 760-mm rainfall (Hyderabad)

	Loamy sand		Sandy loam	
Soil characteristics	Initial	After 5 years	Initial	After 5 years
Percent aggregate (>0.25 m)	62	84	80	87
Mean weight diameter (mm)	3.21	4.19	2.63	3.04
Hydraulic conductivity (cm/hr)	2.84	8.00	4.19	10.53
Bulk density (g/cm^3)	1.78	1.65	1.81	1.69
Soil crusting (Pentrometer reading using $\frac{1}{4}$-inch diameter piston needle)	40	11	35	18

Source: Venkateswarlu (1984b).

IX. Runoff Collection and Recycling

Runoff is inevitable in tropics and subtropics. Different estimates are available (Randhawa and Venkateswarlu, 1980; Verma 1982; Miranda et al., 1983; El Swaify et al., 1985). Broadly speaking, the runoff estimates are:

Amount of rainfall	*Average runoff (%)*
Low	9.7
Medium	25.3
High	38.1

Efforts have been made in storing a part of the runoff in dugout ponds for recycling to the donor area. A limit of 33% of the runoff was attempted to be harvested into these farm ponds. In such a case, the size of the pond generally works out to 250 m^3 ha^{-1} of donor area. This would be adequate to give a 50-mm irrigation once to half the donor area after each fill.

Some aspects that received attention in India include:

1. Identification of critical periods of crops when the effects of moisture stress would be maximum

Table 48. Percent yield reduction due to moisture stress at different crop growth stages (Sand Culture Studies)

	Millets			
	Sorghum		Pearl millet	Foxtail millet
Stage of stress	CSH-1	M-35-1	(HB-3)	(K221-1)
Control (no stress)	0	0	0	0
Crown root initiation	17.2	32.1	25.9	21.3
Bootleaf	65.6	100.0	42.6	74.5
Panicle emergence/flowering	—	66.7	38.0	82.3
Softdough/milk stage	39.3	34.1	60.3	85.4
	Commercial crops			
	Cotton (Krishna)		Safflower (7-13-3)	Sunflower (EC-68415)
Control	0		0	0
Rosette/seeding/square formation	42.4		65.5	24.1
Flower initiation	44.7		92.3	55.6
Peak blooming/boll formation/flowering	32.8		61.3	71.8
Seed setting/boll bursting	28.4		27.2	69.1

Source: Venkateswarlu and Tata Rao (1972); Ramamohan Rao et al. (1977).

2. Sealing the ponds to prevent the percolation losses
3. Efficient use of harvested runoff water

A. Critical Periods

Sand culture studies were conducted by Venkateswarlu and Tata Rao (1972) and Ramamohan Rao et al. (1977). The results (Table 48) show that three stages (namely, crown root initiation, tillering/branching, flowering, and seed selting/boll bursting were critical in reducing yield of different crops.

Several field studies of ICRISAT and AICRPDA also illustrate the point (Kanwar et al., 1982; Vijayalakshmi, 1983). The payoff for an irrigation during these critical periods worked out to about 200 kg cm^{-1} in the case of cereals and 100 kg cm^{-1} for food legumes and oilseeds (Venkateswarlu, 1985).

B. Sealing of Ponds

While the problem of percolation losses is less with deep Vertisols, it is serious in light-textured Alfisols, Inceptisols, and Aridisols. Some efforts have been made in identifying sealants for effectively arresting the percolation losses. Miranda et al. (1983) found just compaction would be adequate in Vertisols, while a 10:1 ratio of soil to cement is needed in Alfisols (Table 49). Black

Table 49. Evaluation of sealants

Research center	Material used	Seepage reduction when compared with control (%)
ICRISAT		
Alfisol	Soil:cement (10:1)	97
	Red soil:black soil (1:2)	90
Vertisol	Na_2CO_3	33
	Red soil:black soil (1:2)	33
Bangalore (Alfisol)	Soil:cement (8:1)	58
Dantiwada (Aridisol)	Plastic lining overlaid by brick	91
	Lime mortar (1:6) with asphalt lining	89
	Cement:sand (1:6)	81
Hyderabad	Plastic sheet overlaid by brick	100
(Alfisol)	Asphalt	87
Ludhiana (Inceptisol)	Bottom polythene	95
Rajkot (Vertisol)	Soil compacted with high bulk density	57
Ranchi	Coal tar (asphalt)	56
(Vertisol)	Clay	
Varanasi	Black polythene	96
(Entisol)	Soil:cement (10:1)	76

Source: Miranda et al. (1983); Vijayalakshmi (1983).

polythene lining in Entisols of Varanasi, bottom polythene in Inceptisol of Hoshiarpur, and plastic sheet overlined with bricks and lime plaster in Alfisols of Hyderabad and in the Aridisols of Dantiwada were found to be effective (Table 49) by Vijayalakshmi (1983). The thickness of the plastic films had been of 400–600 gauge (0.10–0.15 mm thickness).

Natural siltation, of course, also helps in sealing the ponds, but it is for larger ponds (Table 50). It also needs more time.

C. Efficient Use

The harvested water should never be considered for use as a normal irrigation source. It must be used (1) for critical stages of crop growth, and (2) in a most efficient way. Flooding versus furrow irrigation was tested (Table 51) in different centers and it was found that furrow-alternate furrow method would be more effective.

Lifting and conveyance devices have not yet received full attention in runoff recyling. Furthermore, these methods of water harvesting have not, so far, been fully tested on real farm situations.

Table 50. Effectiveness of natural silting in farm ponds

Research center	Year	Seepage as % of 1st year
Rajkot (Vertisol)	1st	100
	3rd	25
	5th	14
Dehra Dun (Inceptisol)	1st	100
	5th	38
	10th	24

Source: Vijayalakshmi (1983).

Table 51. Evaluation of methods of application of runoff water

Crop	Research center (soil type)	Yield of grain ($Mg\ ha^{-1}$) Flooding	Furrow	Alternate furrows
Sorghum	Jhansi (Vertisol)	1.63	1.88	1.98
Sorghum	Bijapur (Vertisol)	—	2.91	2.63
Castor	Anantapur (Alfisol)	0.39	0.36	0.37
Pigeonpea	Anantapur (Alfisol)	0.66	0.71	0.80
Peanut	Rajkot (Vertisol)	—	0.75	0.89
Rice	Bhubaneswar (Alfisol)	0.74	0.80	0.62
Chillies	Hyderabad (Alfisol)	7.33	8.44	10.22

Source: Vijayalakshmi (1983).

X. Stability in Crop Production in Drylands

Broadly speaking, the rainfall determines the cropping pattern. Based on the rainfall patterns, the more efficient crops have been identified by the National Commission on Agriculture (1976). For example:

Rainfall for 3 months (mm/month)	*Efficient crops*
>300	Rice (*Oryza sativa* L.)
200–300	Corn, sorghum
100–200	Sorghum, pearl millet
50–100	*Phaseolus aconitifolius*
<50	Nonproductive

Thus choice of crops would be of importance for any given agro-ecological situation. The existing cropping patterns have been analyzed by the National Commission on Agriculture (1976) in relation to rainfall. They identified 62 rainfall regions in India and also enumerated the existing cropping patterns. Later Singh (1979) delineated India into 60 agro-ecological zones based on soils, rainfall, and temperature.

Yet another important factor is the variability in the rainfall. The probability of deficient rainfall (deficiency in rainfall numerically equal to or greater than 25% of the average rainfall) in India during the southwest monsoon period is as follows:

Meteorological subdivision	*Probability of highly deficient rainfall*
Assam	Very rare; once in 15 years
West Bengal, Madhya Pradesh, Konkan, Coastal Andhra Pradesh, Kerala, Bihar, Orissa	Once in 5 years
South interior Karnataka, east Uttar Pradesh, Vidarbha	Once in 4 years
Gujarat, east Rajasthan, west Uttar Pradesh, Tamil Nadu, Jammu and Kashmir, Rayalaseema and Telengana	Once in 3 years
West Rajasthan	Once in 2.5 years

Axiomatically then, the stability in production also varies (Table 52) over different agro-climatic conditions. The data clearly suggest:

1. The coefficient of variation in productivity increases with decrease in rainfall.
2. In low-rainfall areas (Bellary), such a variation is more with traditional practices.
3. The productivity remains higher with improved management practices.

Table 52. Variability in yield of crops in different agro-climatic regions

Center	Soil type	Average rainfall (mm)	Crop	Average yield (Mg ha^{-1})	SD	CV (%)	Remark
Varanasi	Entisol	1,080	Rice	2.25	0.40	14.6	Monsoon crop
Hoshiarpur	Inceptisol	1,000	Wheat (improved)	3.15	1.01	32.0	Postmonsoon crop
			Wheat (traditional)	1.87	0.55	29.4	Postmonsoon crop
Hyderabad	Alfisol	760	Sorghum (improved)	3.55	1.52	39.0	Monsoon crop
			Sorghum (traditional)	1.92	0.75	39.3	Monsoon crop
Bellary	Vertisol	508	Sorghum (improved)	1.89	0.86	46.0	Postmonsoon crop
			Sorghum (traditional)	0.77	0.48	62.0	Postmonsoon crop
Jodhpur	Aridisol	380	Mungbean	0.93	0.65	65.1	Monsoon crop

Source: Annual Reports of the respective centers.

SD = Standard deviation; CV = coefficient of variation.

Krishnamoorthy et al. (1971), in fact, pointed out that the best way to identify varieties of crop plants for drought tolerance is to identify cultivars with high yield potential, more so when the stress due to lack of moisture is nonspecific.

A. Matching Crops/Varieties with Rainfall

Based on the analysis of rainfall and water storage characteristics of the soils, the assured moisture availability periods could be identified as pointed out earlier. One way, then, to select crop/varieties is to match this period of dependable moisture supply. For instance, Krishnamoorthy et al. (1975) pointed out that the best way to select rice varieties for east Uttar Pradesh was to identify the assured rainfall periods and to choose photoinsensitive varieties.

Some more recent examples further illustrate the point. Sunflower was thought to be an answer to the edible oils problem in India when it was introduced on an extensive scale in 1973–1974. Soon the yields started declining. Putt (1975) felt that the problem was mostly due to poor seed setting, inadequate plant stands, and improper planting patterns. The drive on this crop was almost given up. But today, a little more than 850,000 ha of land is covered with sunflower. A careful analysis showed that up to 800,000 ha are in medium-deep Vertisols and Vertic soils spread over medium-rainfall areas. And all this is because of the intellect of the Indian farmer. As to the scientific reasons, it is clear that sunflower, which has a short duration (90 days), has a grand growth period between 35 and 65 days. So, if there is any stress in this period, the crop is bound to suffer in productivity. And in a medium black soil of medium rainfall with an upper storage limit of 0.15 m^3 per m of soil water, there is an increased possibility of this crop escaping any stress.

Similarly, peanut in the Saurashtra region suffers from lack of water at the later stages of crop growth. The rainy season in the Rajkot, Amerali, and Gogha regions is about 12–13 weeks (Virmani et al., 1982). The duration of the conventional bunch varieties is about 110 days. But when the JL-24 cultivar of 100 days was identified, it found favor with the environment in this peanut area. Thus it is important to match duration with the growing season.

In fact, earlier efforts at Bellary pointed out the need to replace the existing tall, long-duration varieties of sorghum with shorter duration cultivars, as the growing season in Bellary is only 100 days (Krishnamoorthy et al., 1977). Similarly, cotton (*Gossypium hirsutum* L.) in the Deccan as a postrainy crop has yielded very poorly and could be replaced with hardy safflower (*Carthamus tinctorius* L.). Below the Vindhyan region, where the temperatures rarely drop below 15°C and where winter rains are almost absent, wheat needs to be replaced with crops like chickpea and safflower.

Based on such data as well as that of rainfall and soil, zonality for cropping is being recognized in the Indian context. Some examples are given below (Venkateswarlu, 1985).

Crop	*Potential areas*
Cereals	
Rice	Medium lands in a catenary situation in northeast India
Wheat	Above Vindhyan region with assured winter rains
Sorghum	Shallow to medium Vertisols and vertic soils of Andhra Pradesh, Karnataka, and Maharashtra as a rainy crop
	Deep Vertisol of Tamil Nadu, Deccan postrainy region, and east Telengana
Oilseeds	
Peanut	Continuing efforts in the existing areas, new thrust being given in east Madhya Pradesh, South Bihar plateau, northwest Orissa, and west Uttar Pradesh
Rape/Mustard	Nonfrost, winter rain areas of north India
Safflower	As a second crop in the Entisols of north India; deep Vertisols of Madhya Pradesh and east Rajasthan
	Sole crop in Vertisols of Tamil Nadu, Entisols of Agra region, and Sierozems of Haryana
Sunflower	Medium-deep Vertisols of the country with medium rainfall
Pigeonpea	Late rainy crop in Vidarbha region
	August-September crop in Vertisols of Andhra Pradesh
	Sole crop in the Deccan region
	Second crop in the plains of Bihar
Chickpea	Second crop in high-rainfall Vertisols below Vindhyan region and in Entisols of north India after rice/pearl millet
	Sole crop in low-rainfall Deccan region and Sierozems in northwest India

Once the most suitable crops are identified, increased productivity is possible with better crop management practices. However, over a given period the variation in yields would be more as the rainfall decreases. This then calls for capitalization on good rainfall years.

When crops are grown in the rainy season, the "breaks" in the monsoon are important. If the breaks occur during the critical stages of the growth of a given crop, the effects will be more devastating. However, if the assured rainy periods are identified and dates of sowing are accordingly adjusted to avoid these critical phenological stages, the effects of "breaks" in the monsoon can be minimized. For instance, in sorghum, the flowering and seed-filling stages (8–10 weeks after sowing) are important stages. And in the semiarid Alfisol region of Hyderabad, the sorghum (CSH-6) yields and rainfall were relative by

$$Y = 46.32 - 2.22\,W - 1.55\,\mathrm{P}_1 - 4.7\,\mathrm{P}_2 + 15.46\,\mathrm{P}_3$$

where W is seeding week, and P_1, P_2, P_3 are log-transformed rainfall (mm) during 4–5, 6–7, and 8–10 weeks of crop growth, respectively.

The rainfall during 8–10 weeks of crop growth was highly correlated ($R^2 = 0.95$, significant at 1% level) with the yield and thus this is a critical period for the sorghum crop (Bhaskara Rao and Vijayalakshmi, 1985). The average weekly rainfall for this period of 8–10 weeks after sowing was worked out by them (Table 53) and the probabilities of obtaining good, average, and poor rainfall have also been worked out. The data imply that a good crop of sorghum is likely in only 23% of the years in this region and that the crop would be poor in 17% of the years.

Similar probabilities have been worked out for castor (*Ricinus communis* L.), a deep-rooted crop in the same region (Bhaskara Rao and Vijayalakshmi, 1985). The interrelation worked out as follows:

$$Y = -0.15 + 0.0162\ P_1 + 0.0118\ P_2 + 0.0690\ P_3$$

where Y is yield, and P_1, P_2, P_3 are the log-transformed rainfall (mm) values during 31–60, 61–90, and 91–120 days of crop growth, respectively. The regression coefficients indicate that the 91–120-day period is more important for the castor crop in the Semi-arid tropic (SAT) Alfisols. The average weekly rainfall during this period along with probabilities for good, average, and poor events are given in Table 54. Thus, in the region, castor has a 44% probability of a good crop compared to only 23% for sorghum.

Table 53. Probabilities of performance of sorghum and corresponding rainfall during weeks 8–10 (flower and grain-filling period) in Alfisols of Hyderabad region

Class	Probability	Average weekly rainfall (mm) during weeks 8–10
Good	0.23	62.9
Average	0.60	26.4
Poor	0.17	6.1

Source: Bhaskara Rao and Vijayalakshmi (1985).

Table 54. Probabilities of success of castor in Alfisols of Hyderabad region

Class	Probability	Average weekly rainfall in 91–120-day period (mm)
Good	0.44	55.8
Average	0.36	27.8
Poor	0.20	12.8

Source: Bhaskara Rao and Vijayalakshmi (1985).

Sorghum, as a principal crop, has a problem of stability. In order to increase stability, intercropping with pigeonpea is suggested. Rao and Willey (1980) analyzed stability in yield of sole sorghum and pigeonpea and related it to intercropping 51 experiments conducted all over India. The coefficient of variation was only 39% with intercropping, compared to 49% with sorghum and 44% with pigeonpea.

XI. Agro-Technology Transfer

Ikawa (1979) reported earlier on agro-technology transfer through use of Bench Mark Soils. The basic assumptions made were (1) that the soils in the same category have similar management problems, and (2) that the soil family provides the link between soil classification and land capability.

Swindale (1978) pointed out that although an adequate national input is being made into soil survey and classification, this is not so for soil survey interpretation and land use classification. An understanding of soil potential and limitation is essential to good land use planning for good presentation and economic development and for wise use of soils. Soil survey interpretation and land classification, like soil management, are approaches toward a holistic understanding of soils.

Murthy (1978) discussed land use planning for the Bangalore district of Karnataka. He identified a 33-ha site with six soil mapping units as follows:

1. Hoshote sandy loam, very deep, on 0–1% slope, moderately eroded .... 0.51 ha
2. Hoshote sandy clay loam, very deep, on 0–1% slope, moderately eroded .... 9.78 ha
3. Hoshote sandy loam, very deep on 0–3% slope, slightly eroded .. 6.59 ha
4. Hoshote sandy clay loam, very deep, on 1–3% slope, moderately eroded .... 10.12 ha
5. Hoshote clay loam, very deep, on 1–3% slope, moderately eroded .... 4.92 ha
6. Miscellaneous (rock out crops, etc.).......... 0.1 ha

Crops like finger millet (*Eleusine coracana* (L.) Gaertn.) were utilized as the principal crop and the yields over 2 years were as listed in Table 55.

Table 55. Yield of crops improved land use system

	Yield (Mg ha^{-1})	
Crop	Conventional system	Improved system
Finger millet	0.95	2.95
Hybrid corn	1.55	3.05
Sunflower	5.00	8.10
Cowpea	4.00	7.00

In this preliminary test it was found that, based on the depth of soil and texture, the soil moisture that could be stored was known. Accordingly, the expected yields could be estimated. Double-cropping in deep and heavier soils could be attempted.

Shankaranarayana et al. (1980) worked on the arid and semiarid regions of the north Indian plains using soil maps refined with information on textural classes. In the Hisar district, they found that in dry regions, sandy soils and coarse-textured soils can support only short-duration rainy season crops. And in the wet regions of Ganganagar, textural classes helped in determining the cropping pattern. In finer textured soils, long-duration crops could be taken up, while medium-range soils could be cropped to pearl millet and *Eruca sativa*.

In both instances an intimate knowledge of the site helped in proper crop planning. When these finer points are brought out on a soil series basis, it should be possible to extend the improved practices from known to unknown areas.

Shankaranarayana et al. (1980) studied the effect of texture as reflected in the soil series on the yield of irrigated wheat, mustard, and chickpea. They found that coarse loamy soils gave better yield of wheat as compared to fine sandy loams. Fine loamy soils and areas gave medium performance. Such differences were absent with the other crops. The authors called for high level of inputs for increased yields in fine sandy loam soils.

Bhattacharjee and Landey (1984) identified minimum data sets needed for agro-technology transfer. They include:

1. Soil depth
2. Effective depth of root ramification zone
3. Particle size distribution, particularly of fine clay fraction ($<0.1\ m^{-1}$)
4. Cation exchange capacity (CEC) and saturating cations
5. Bulk density and coefficient of linear extensibility (COLE) values
6. Soil temperatures at different moisture tensions
7. Saturated hydraulic conductivity and depth of water table
8. Available moisture retentivity
9. Level of available nutrients
10. Weekly data of climatic components influencing crop production
11. Socioeconomic status of the farmer

Ghodake et al. (1985), working on qualitative evaluation of agro-technology transfer potential on associated families of Vertisols, found the scope of transferring technology in crop production in one soil series to another. They also found that a few of the soil series gave similar responses in crop production. But lack of response to fertilizer nitrogen is unique in this study.

In this study, they found that this transfer is possible only at a higher level of management, which constitutes about 30% in their sample study. Even for this endeavor, the taxonomic characteristics that are the inherent properties of the soils become important.

To sum up, agro-technology transfer based on soil classification and land capability is a new area of research and is gaining popularity in India. The requirements are a detailed soil map (4 inches = 1 mile) and supporting data from well-designed simple experiments taking soil series as a basis.

Acknowledgment

I thank Dr. N.S. Randhawa, Director-General, Indian Council of Agricultural Research, New Delhi—110 001, India for encouraging me to take up this assignment.

References

Anderson, D.T. 1980. Soil management. *Andhra Agric. J.* 27:215–225.

Baver, L.D., W.H. Gardner, and W.R. Gardner. 1977. *Soil Physics*, 4th Ed., Wiley, New York.

Bharadwaj, K.K.R. 1981. Potentials and problems in the recyling of farm and city wastes on the land. In: *Recycling Residue of Agriculture and Industry*. M.S. Kalra (ed.), Punjab Agricultural University, Ludhiana, India, pp. 57–76.

Bhaskara Rao, U.M., and K. Vijayalakshmi. 1985. *Crops and Cropping Systems for Semi-Arid Red Soils of Telengana Region*. Central Research Institute for Dryland Agriculture, Hyderabad, India. Research Bulletin 4.

Bhattacharjee, J.C., and R.J. Landey. 1984. Minimum data sets for agro-technology transfer on Vertisols and associated soils of India. *Proceedings of the International Symposium on Minimum Data Sets for Agroforestry Transfer*., March 21–26, 1983. ICRISAT Centre, Patancheru, pp. 11–23.

Brewbaker, J., and E.M. Hilton. 1979. *Leucaena versatile* tropical tree legumes. In: *New Agric. Crops*, AAAS, Selected Symposia, West View Press, Boulder, Colorado, pp. 207–259.

Burnett, E. 1985. Conservation production system for dryland agriculture in the United States. *Proceedings of the Workshop on Soil, Water and Crop Management Systems for Rainfed Agriculture in Northeast Thailand*, February 25–March 1, 1985. University of Khon Kaen, Khon, Kaen, Thailand, in press.

Chaudhury, T.N., and B.S. Sandhu. 1982. Soil temperature and plant growth. *Review of Soil Research in India*, Part II. 12th International Congress on Soil Science, New Delhi, pp. 60–76.

Cocheme, J., and P. Franquin. 1967. *A Study of the Semi-Arid Area of the Sahara in West Africa*. FAO/UNESCO/WMO/Inter Agency Project, pp. 1117–1129.

De, R., and S.P. Singh. 1983. Cropping systems under adverse weather conditions. *Ferti. News* 29(2):22–26.

Dhruvanarayana, V.V., and Ram Babu. 1983. Estimates of soil erosion in India. *J. Irrig. Drainage Eng.* 109:419–436.

Dixit, L.A., B.R. Hegde, and T.C. Chennappa. 1981. *Soil and Moisture Conservation Practices in Red Soils*. Panel discussion on soil and water conservation in red and black soils. Research Centre, Central Soil and Water Conservation Research and Training Institute, Bellary and University of Agricultural Sciences, Bangalore, pp. 28–32.

El Swaify, S.A., P. Pathak, J.J. Rego, and S. Singh. 1985. Soil management for optimized productivity under rainfed conditions in the semi-arid tropics. In: B.A. Stewart (ed.), *Advances in Soil Science*, Vol. 1. Springer-Verlag, New York, pp. 1–64.

Finck, A. and J. Venkateswarlu 1982. Chemical properties and fertility management of Vertisols. In Vertisols and Rice Soils of the Tropics, Symposia Vol. II. 12th International Congress on Soil Science, February 8–16, New Delhi 61–79.

Gardner, E.A., R.J. Shaw, G.D. Smith, and K.J. Coughlan. 1983. Plant availability capacity: Concept, measurement and prediction. *In*: J.W. McGarity, E.W. Houltz, and H.B. So (eds.), *Properties and Utilization of Cracking Clay Soils. Reviews in Rural Science*, No. 5, University of New England, Armidale, N.S. Wales.

Ghodake, R.D., L.D. Swindale, K.V. Subba Rao, and J.C. Bhattacharjee. 1985. *Quantitative Evaluation of Agro-Technology Transfer Potential on Associated Families of the Vertisols.* Collaborative project between ICRISAT, ICAR, NBSS, and LUP on Bench March Soils Network for Agro-technology Transfer (mimeographed).

Guenzi, W.D., T.M. McCathy, and F.A. Norstadt. 1967. Presence and persistence of phytotoxic substances in wheat, oat, corn and sorghum residues. *Agron. J.* 59: 163–165.

Gupta, J.P., and G.N. Gupta. 1981. A note on wind erosion from a cultivated field in West Rajasthan. *J. Indian Soc. Sci.* 29: 278.

Gupta, R.P., S. Kumar, and T. Singh. 1984. *Consolidated Report*; *Soil Management to Increase Crop Production.* IARI, New Delhi, p. 104.

Hadimani, A.S., B.R. Hegde, and T. Satyanarayana, 1982. Management of red soils. *Review of Soil Research in India*, Part II. 12th International Congress on Soil Science February 8–16, 1982. New Delhi, India, pp. 689–702.

Hargreaves, G.H. 1975. *Water Requirements Manual for Irrigated Crops and Rainfed Agriculture.* EMBRAPA and Utach State University Publication 75-D 158.

Hudson, N. 1971. *Soil Conservation.* BT Batsford Limited, London.

Ikawa, H. 1979. *The Bench Mark Soils Project of Hawaii and Puerto Rico: Research a Method for Agro-technology Transfer in Tropics.* Report of an Expert Consultation on Land Evaluation Guidelines for Rainfed Agriculture, Rome, December 1979, pp. 108–113.

International Crops Research Institute for the Semiarid Tropics (ICRISAT). 1984. *Farming Systems.* Annual Report 1983, pp. 233–307.

Joffe, J.S. 1955. Green manuring viewed by a pedologist. *Adv. Agron.* 7: 142–187.

Kang, B.T., G.F. Wilson, and L. Sipkens. 1981. Alley cropping maize (*Zee mays* L.) and Leucaena (*Leucaena leucocephala* Lam) in southern Nigeria. *Plant Soil* 63: 165–179.

Kanwar, J.S., J. Kampen, and S.M. Virmani. 1982. *Management of Vertisols for Maximizing Crop Production—ICRISAT Experience. Vertisols and Rice Soils of the Tropics.* Symposia Papers II, 12th International Congress on Soil Science, February 8–16, 1982. New Delhi, India, pp. 94–120.

Kanitkar, N.V., S.S. Sirur, and D.H. Gokhale. 1968. *Dry Farming in India.* Indian Council of Agricultural Research, New Delhi.

Krishnamoorthy C., S.L. Chowdhury, D.T. Anderson, and R.D. Dryden. 1977. Cropping systems for maximizing production under semi-arid conditions. *Proceedings of the Second FAO/SIDA Seminar on Field Food Crops in Africa and the Near East*, September 18–October 5, Lahore (Pakistan) (mimeographed).

Krishnamoorthy C., S. Freyman, and E.D. Spratt. 1975. *Concept of Non-specific Stress.* Symposium on Crop Plant Response to Environmental Stresses, Almora, July 1975 (mimeographed).

Krishnamoorthy C., S.V.S. Shastry, and W.M. Freeman. 1971. *Breeding Rice for Drought Tolerance and Salinity*. Oryza 8:47–56.
Krishnamoorthy C., and J. Venkateswarlu. 1976. Nutrient availability concepts and relationships. In: J.S. Kanwar (ed.), *Soil Fertility—Theory and Practice*. Indian Council of Agricultural Research, New Delhi, pp. 65–92.
Kumara Rao, J.V., P.J. Dart, and P.V.S.S. Sastry. 1983. Residual effect of pigeonpea (*Cajanus cajan*) on yield and nitrogen response of maize. *Exp. Agric.* 19:131–141.
Lockeretz, W., G. Shearer, and D.H. Kohl. 1981. Organic farming in corn belt. Science 211:540–456.
Mallick, S., and Y. Nagaraja Rao, and G. Singh. 1978. Soil physical environment and root distribution in irrigated maize and wheat under different soil management practices. *Indian J. Agron.* 23:295–300.
Miranda, S.M., P. Pathak, and K.L. Srivastava. 1983. Run-off management on small agricultural watersheds. ICRISAT experience in dryland agricultural research in India. In: R.P. Singh and V. Bala Subramanian (eds.), *Thrust in the Eighties*. AICRPDA, Hyderabad, pp. 88–97.
Murthy, R.S. 1978. Land-use planning in Karnataka in Soil Resource Data for Agricultural Development. Ed L.D. Swindale, Hawaii Agrl. Exptl. Sta., College of Tropical Agric., Univ. Hawaii, Honolulu, Hawaii, 104–129.
Murthy, R.S., L.R. Hirekarur, S.B. Deshpande, B. Venkat Rao, and K.S. Shankaranarayana. 1981. *Bench March Soils of India*. National Bureau of Soil Survey and Land Use Planning (ICAR), p. 161.
National Commission on Agriculture. 1976. *Consolidated Report*. National Commission on Agriculture, New Delhi.
Oakes, A.J. 1968. *Leucaena leucocephala*: Description, culture, utilization. *Adv. Front. Plant Sci.* 29:1–114.
Pratap, N., B. Verma, and A.K. Singhal. 1980. Nitrogen economics through intercropping of pigeonpea in rainfed sorghum. *Indian J. Agron.* 25:190–196.
Prihar, S.S., S.T. Sandhu, K.L. Khera, and S.K. Jalota. 1978. Water use and yield of winter wheat in northern India by timing last irrigation. *Irrig. Sci.* 1:39–45.
Prihar, S.S., R. Singh, N. Singh, and K.S. Sandhu. 1979. Effect of mulching on previous crop or fallow on dryland maize and wheat. *Exp. Agric.* 15:129–135.
Putt, E.D. 1975. *Problems of the Sunflower Crop in India*. All India Co-ordinated Research Project for Dryland Agriculture, Hyderabadm (mimeographed).
Ramamohan Rao, M.S. 1981. *Moisture Conservation for Higher Crop Yields in Black Soils*. Panel discussion on soil and water conservation in red and black soils. Central Soil and Water Conservation Research and Training Institute, Research Centre, Bellary and University of Agricultural Sciences, Bangalore, pp. 48–53.
Ramamohan Rao, M.S., and N. Seshachalam. 1976. Improvement of intake rate in problem black soils. *Mysore J. Agric. Sci.* 10:52–58.
Ramamohan Rao, M.S., N. Seshachalam, V. Ranga Rao, and M. Ramachandran. 1977. Critical stages for moisture stress in different crops. *Mysore J. Agric. Sci.* 11:494–500.
Randhawa, N.S., and J. Venkateswarlu. 1980. Indian experience in the semi-arid tropics—Prospects and retrospects. ICRISAT (International Crops Research Institute for the Semi-arid Tropics). *Proceedings of the International Symposium on Development and Transfer of Technology for Rainfed Agriculture and SAT farmers*, August 28–September 1, 1979. Patancheru, A.P., India, pp. 207–220.
Rao, M.R., and R.W. Willey. 1980. Evaluation of yield stability in intercropping. Studies on sorghum/pigeonpea. *Exp. Agric.* 16:105–116.

Reddy, D.S., and C. Dakshinamurthy. 1971. Root growth and soil structures under different tillage operations and uniform fertilizer application. *Indian J. Agric. Sci.* 40:413–422.

Russell, M.B. 1978. Profile moisture dynamics of soil in Vertisols and Alfisols. ICRISAT (International Crops Research Institute for the Semi-arid Tropics). *Proceedings of the International Workshop on Agroclimatological Research Needs of the Semi-arid Tropics.* Novermber 22–24, 1978. Hyderabad, India, pp. 75–87.

Sandhu, B.S., S.S. Prihar, K.L. Khera, and K.L. Sandhu. 1978. Scheduling irrigation to chickpea. *Indian J. Agric. Sci.* 48:486–491.

Shankaranarayana, H.S., L.R. Hirekerur, C.M. Mathur, and R.S. Murthy. 1980. Soil maps for land evaluation with reference to crop yields. *Ann. Arid Zone* 19:185–187.

Shriniwas, S., V. Kameswara Rao, and D.S. Rajput. 1981. *Soil Erosion and Its Control in the Red Soil Areas of Telengana.* Panel discussion on soil and water conservation red and black soils. Research Centre, Central Soil and Water Conservation Research and Training Institute, Bellary and University of Agricultural Sciences, Bangalore.

Singh, A. 1964. Soil compaction effects on the growth of sugarcane roots. *Sci. Cult.* 30:606–607.

Singh, H.P. 1982. Management of desert soils. *Review of Soil Research in India,* Part II. 12th International Congress on Soil Science, February 8–16, 1982, New Delhi, India, pp. 676–688.

Singh, R., Y. Singh, S.S. Prhar, and P. Singh. 1975. Effect of N fertilization on yield and water use efficiency of dryland winter wheat as affected by stored water and rainfall. *Agron. J.* 67:599–603.

Singh, R.P., and S.K. Das. 1984. *Nitrogen Management in Cropping Systems with Particular Reference to Rainfed Lands of India.* Part A of Project Bulletin No. 8, AICRPDA, Hyderabad.

Singh, R.P., H.P. Singh, H.S. Daulay, and K.C. Singh. 1981. Fertilization of rainfed greengram-pearl millet sequence. *Indian J. Agric. Sci.* 51:498–503.

Singh, S.N. 1978. *Short Period Soil Water Availability at Dryland Research Centres.* All India Co-ordinated Research Project for Dryland Agriculture, Hyderabad, Ag. Climate, p. 5 (mimeographed).

Singh, S.N. 1979. *Agro-ecological Zones of India.* All India Co-ordinated Research Project for Dryland Agriculture, Hyderabad.

Smith, G.D., D.F. Yule, and K.J. Caughlan. 1984. *Soil Physical Factors in Crop Production on Vertisols in Queensland, Australia.* ACIAR (Australian Council of International Workshop on Soils), September 12–16, 1983, pp. 87–104.

Smith, G.W. 1942. Sanborn field "Fifty years of field experiments with crop rotation, manures and fertilizers." *Missouri Agric. Exp. Sta. Bull.* 458.

Soil Survey Staff. 1978. Soil taxonomy: A basic system of soil classification for making and interpreting soil surveys. *Agriculture Hand Book No. 436.* USDA, SCA, U.S. Government Printing Office, Washington, DC.

Stewart, B.A., and E. Burnett. 1985. Water conservation technology in rainfed and dryland agriculture. *Proceedings of the International Conference on Food and Water,* May 27–30, 1985. College Station, Texas.

Subba Rao, N.S., K.B.V.R. Talik, C.S. Singh, and R.C. Gautam. 1982. Crop yield and nitrogen content of pearl millet (*Pennisetum americanum*) in response to *Azospirillum brasilonse* in biological nitrogen fixation. *Proceedings of the National Symposium on IARI by Bhaba Atomic Research Centre,* Bombay, pp. 507–576.

Swindale, L.D. 1978. A soil research network through tropical soil families. In: L.D. Swindale (ed.), *Soil-Resource Data for Agricultural Development*. Hawaii Agricultural Experimental Station College of Tropical Agriculture, University of Hawaii, Honolulu, Hawaii, pp. 210–218.

Tejwani, K.G., and S.P. Bharadwaj. 1982. Soil and water conservation research. *Review of Soil Research in India*, Part I. 12th International Congress on Soil Science, February 8–16, 1982. New Delhi, India, pp. 608–621.

Umrani, N.K. 1980. Research achievements with special reference to fertilizer use in drylands—problems encountered, suggestions field experience to solve them in Mahavashtra, Proc. FAI Group Discussion on Fertilizer use in Drylands. New Delhi, pp. 55–61.

Venkat Reddy, N., and R. Hampaiah. 1982. Alternate research in India. In: G. Friesen, J. Venkateswarlu, B.K. Rastogi, and V. Bala Subramanian (eds.), *A Decade of Dryland Agricultural Research in India* 1971–80. AICRPDA, Hyderabad, pp. 74–89.

Venkatasubbaiah, V., J. Venkateswarlu, and V.V.K. Sastry. 1976. Potassium supplying power of black soils of West Godavari, Andhra Pradesh, *Bull. Indian Soc. Sci.* 10:219—226.

Venkateswarlu, J. 1980. Improved dryland farming systems in India. *Andhra Agric. J.* 27:133–147.

Venkateswarlu, J. 1981. Maximization of crop production in drylands. *Indian J. Soil Cons.* 9:124–140.

Venkateswarlu, J. 1984a. *Ecological Crises in Agro-ecosystem.* India's Environment, crises and responds, J. Bandyopadhyay, N.D. Jayal, V. Schoettli and Chatrapati Singh (eds.), Natraj Publishers, Dehra Dum pp. 91–109.

Venkateswarlu, J. 1984b. *Nutrient Management in Semi-arid Red Soils.* Part II, Project Bulletin, No. 8, AICRPDA, Hyderabad.

Venkateswarlu, J. 1985. *Dryland Farming.* Central Research Institute for Dryland Agriculture (CRIDA), Hyderabad Extension Bulletin No. 1.

Venkateswarlu, J. 1986. *Rainfed Farming and Phosphate Fertilizer Management.* Proc. Second Regional Seminar on Crop Production Techniques and Fertilizer Management in Rainfed Agriculture in Southern Asia, Jan 22–25 1986, New Delhi. International World Phosphate Institute, pp. 109–124.

Venkateswarlu, J., and R. Singh. 1982. Crop responses to applied nutrients under limited water conditions. *Review of Soil Research in India*, Part I. 12th International Congress on Soil Science, February 8–16, 1982. New Delhi, India, pp. 442–452.

Venkateswarlu, J., and B.V. Subbiah. 1964. The effect of phosphate additions on fertilizer sulphur availability. *Proceedings of the Symposium on Nuclear and Radiation Chemistry*, 1964. Bombay, pp. 101–104.

Venkateswarlu, J., and U. Tata Rao. 1972. A screening technique for drought tolerance of sorghum varieties. In: N.G.P. Rao and L.R. House (eds.), *Sorghum in the Seventies.* Oxford & IBH Publishing Co., New Delhi, India, pp. 540–544.

Venkateswarlu, J., K.P.R. Vittal, and S.K. Das. 1986. Evaluation of two land configurations in production of pearl millet and castorbean on shallow red soils of semi-arid tropics. Andhra Pradesh, India, International Agrophysics 2:145–152.

Verma, G.P. 1982. Land and water management in semi-arid rainfed areas. *Review of Soil Research in India*, Part II. 12th International Congress on Soil Science, February 8–16, 1982. New Delhi, India, pp. 666–675.

Vijayalakshmi, K. 1982. Water harvesting and reuse. In: G.H. Friesen, J. Venkateswarlu, B.K. Rastogi, and V. Bala Subramnian (eds.), *A Decade of Dryland Agricultural Research in India 1971–80*. AICRPDA, Hyderabad, pp. 103–119.

Vijayalakshmi, K., 1983. Runoff management. In: R.P. Singh and V. Bala Subramanian (eds.), *Dryland Agricultural Research in India: Thrust in the Eighties*. All India Co-ordinated Research Project for Dryland Agriculture, Hyderabad, India, pp. 78–87.

Virmani, S.M., M.V.K. Sivakumar, and S.J. Reddy. 1978. Climatological features of the SAT in relation to the farming systems research programme. ICRISAT (International Crops Research Institute for the Semi-arid Tropics). *Proceedings of the International Workshop on Agro-climatological Research Needs of the Semi-arid Tropics*, November 22–24, 1978. Hyderabad, India, pp. 5–16.

Virmani, S.M., M.V.K. Sivakumar, and S.J. Reddy. 1982. *Rainfall Probability Estimates for Selected Location of Semi-arid India*, Research Bulletin No. 1.

Vittal, K.P.R., K. Vijayalakhmi, and U.M.B. Rao. 1983. Effect of deep tillage on dryland crop production in red soils of India. *Soil Tillage Res.* 3:377–384.

Wani, S.P., S. Chandrapalaiah, and P.J. Dart. 1985. Response of pearl millet cultivars to inoculation with nitrogen fixing bacteria. *Exp. Agri.* 21:175–182.

Index

MIX
Papier aus verantwortungsvollen Quellen
Paper from responsible sources
FSC® C105338

If you have any concerns about our products,
you can contact us on
ProductSafety@springernature.com

In case Publisher is established outside the EU,
the EU authorized representative is:
Springer Nature Customer Service Center GmbH
Europaplatz 3, 69115 Heidelberg, Germany

Printed by Libri Plureos GmbH
in Hamburg, Germany